Red Beans, Rice, and Tabouli

The Best Journally Speakings from 25 years of OGJ

Red Beans, Rice, and Tabouli

The Best Journally Speakings from 25 years of OGJ

Selected and edited by

Leo R. Aalund
John L. Kennedy

PennWell
PUBLISHING COMPANY
TULSA, OKLAHOMA

PennWell Publishing Co.
1421 South Sheridan
Tulsa, Oklahoma 74112

Red beans, rice, and tabouli: The best Journally Speakings from 25 years of OGJ/(edited by) Leo R. Aalund, John L. Kennedy.
p. cm.
ISBN 0-87814-648-2
1. Petroleum—Miscellanea. 2. Petroleum industry and trade — Miscellanea. I. Aalund, Leo R. II. Kennedy, John L. III. Oil & Gas Journal.
TN864.R43 1996
665.5—dc20

96-41195
CIP

Printed in the United States of America
1 2 3 4 5 00 99 98 97 96

Dedicated to Oil & Gas Journal editors,

past and present.

INTRODUCTION

The title of this book comes from one of the more than a thousand Journally Speaking columns written for the Oil & Gas Journal by its editors during the past 25 years. These columns were the place Journal editors unbent a bit from the serious business of covering international oil news and technology and took a look at the world they moved and worked in. This selection of 278 of these from the years 1970 through 1995 was compiled and edited by two veteran Journal editors who have been along for the entire ride.

It was a turbulent one; years of war and peace, oil glut and shortage, and restructuring; years often punctuated with generous portions of stupidity. This quarter century also saw the rise of mighty state owned oil companies and the first Earth Day, which symbolized an environmental storm that spawned sweeping laws, new technologies, and adversaries.

Because Journal editors were influenced by all this and more their columns reflect the tenor of the times, and present history from a different perspective. They were sometimes serious, humorous, or outrageous—but always interesting. The columns had a connection to the oil industry, admittedly sometimes tenuous. A few were controversial.

Former OGJ editorial writer Henry Ralph, the "father" of these columns, created the legendary Churly McCaustic, the nemesis of incremental-barrel refiners, marketers bearing gimmicks, and "rock hounds".

Henry once got a letter he would quote with enjoyment:

"Your backbone is as yellow as the spine on your magazine."

He could take it as well as dish it out.

Henry, and some of our former colleagues whose work appears in this book, have died. Others have retired or moved on to other jobs. But a number are still very active at the Oil & Gas Journal and young enough to be writing columns 25 years from now. Also still at work are two staffers, Stacy Scott and Kim Jordan, who provided valuable assistance in the preparation of this book, and Robert Lair. Bob gets the credit for layout and cover design.

The first column in this book, "Requiem for a Pismo Clam" was written just a few months before the beginning of the past quarter century. It was chosen, however, because it sets the stage for the next 25 years.

We hope our readers, whether their favorite dishes are Louisiana red beans and rice, Middle Eastern tabouli, Asian sushi, or North Sea herring, will relive exciting times in this book, and come away with a laugh or two and greater understanding of the oil world and the magazine that covers it.

Leo R. Aalund, Managing Editor — Technology
John L. Kennedy, Editor in Chief
Oil & Gas Journal

Houston, Summer 1996

TABLE OF CONTENTS

1974

1975

Requiem for a Pismo clam

Dale Basye **April 14, 1969**

On the shores of Santa Barbara north of Santa Monica Bay, a Pismo clam lay dying under the midday sun.

Let's call the little fellow Clyde.

Now, Clyde was no ordinary run-of-the-mill Pismo clam. He was unique. Clyde, you see, wasn't a very bright Californian. He'd gotten himself messed up with some free-floating crude oil in the Santa Barbara Channel. Apparently all the other little sea creatures had been smarter—or sturdier.

Clyde was the only form of small marine life that biologists could find done in by the slick after the channel oil leak was 3 weeks old. And the scientists had scoured 10 miles of the worst-hit areas of the beaches.

Strangely, Clyde's death would have gone unnoticed had he not been spotted by one of the biologists at a tide pool.

The scientist stooped down, picked Clyde up—and that did it.

Had a sea gull glided down and swallowed Clyde in a single gluttonous gulp, would anybody have mourned his fate?

Not on your surfboard!

Or suppose Clyde had been swooped up by a fisherman's net and sold to a fancy Santa Barbara restaurant.

Do you think the diners would have shed a tear as they put Clyde away as an appetizer for a four-course sea-food dinner?

You can bet your sliding hillside home they wouldn't.

Or suppose Clyde had been a Vietnamese clam and got himself clobbered by a Cong mortar while lolling on a Saigon beach, along with a few similarly unlucky human types.

Would anyone plant him in a hole in the sand and blow taps?

Not even a Berkeley peace-rioter.

Had the slick not hit the beach, Clyde's death would have gone about as unnoticed as a newspaper report of another fatality on the San Bernadino freeway. Or a hippie folding his tent from an overdose of drugs.

But Clyde will go down in history. Like John Paul Jones and Lana Turner, he was in the right place at the right time.

And there was much wailing and gnashing of teeth over his passing.

The oil industry triumphs

John Kennedy **March 16, 1970**

If you think Hollywood isn't aware of the problems involved in offshore drilling, you're wrong. The other night the TV schedule billed the late movie as "shrimp fisherman fights the offshore oil drillers." It sounded interesting so I tuned in.

Seems as though, back in '46, Ole Jimmy Stewart was just out of the Navy. He had a platform design that would withstand any storm in the Gulf of Mexico, and he wanted to build one and try it out. Dan Duryea didn't have any pressing engagements at that time, so he was going to be Ole Jimmy's toolpusher. Another part of the story involved an oilman who had a lease in the gulf about to expire. He was a wildcatter at heart. But as president of a company, he was currently very much inhibited by a stodgy board and bespectacled (naturally) accountant.

The problem was to drill a well on the lease before it expired—I think they had about 90 days to erect a platform and drill the well. Got the picture? As you might expect there were obstacles—bad feelings of the shrimp fishermen, the fun-loving nature of the tool-pusher, a hurricane, and a few love conflicts.

The first major crisis came when the fishermen saw the seismic boat shooting up the gulf. It looked bad, but Ole Jimmy was through shooting by the time the fishermen arrived from the governor's office with an injunction. (The governor grants injunctions? So I learned something).

Erection of the platform went pretty well, but before drilling could begin, the hurricane came. The storm was just the thing Ole Jimmy wanted to prove his design. It also brought a couple of problems—the girl in love with Ole Jimmy arrived on the platform just in time to get stranded by the storm (naturally again). And the young man whose girl the toolpusher had stolen arrived with enough dynamite to blow up the structure.

Then things got worse. The well began to flow salt water and when Ole Jimmy called for the BOP to be closed, the handle was stuck. That was solved by using the manual valve handwheels. But what really burned Ole Jimmy was that the toolpusher had sneaked off to shore to get married and missed the whole crisis!

As if that wasn't enough, the company president showed up with the news that money had run out with 8 days to go until the lease expired. Following a fight between Ole Jimmy and the toolpusher, the company president and Ole Jimmy vowed to finish the well themselves. I sure wanted to see that, but I didn't get to. Because a few minutes later, the pusher and the crew agreed to work without pay to see the well through to completion.

And guess what? The picture ended with oil spewing high above the crown block and Ole Jimmy and the toolpusher doing the traditional oil-soaked dance on the rig floor.

They sure had me worried there for awhile.

I didn't say that!!

Frank J. Gardner **April 13, 1970**

You've doubtless heard the old saying, "It loses something in the translation."

Well, we're convinced!

Our man who covers Venezuelan events for us has a sharp eye, and in one of the Caracas newspapers he spotted this headline: VENEZUELA COMPLAINT 'INSOLENCE.' He promptly mailed us the clipping.

The item read thusly: "A columnist classified the Venezuelan complaint that Canada is enjoying preferential treatment to the U.S. petroleum market as one of 'continual insolence.'

"Frank J. Gardner, in an article of the Industrial Weekly Report, stated that the Canadian Embassy in Caracas has released information pertaining to the 'truthful version of the petroleum situation'."

Well, now . . . we don't recall ever having used the word "insolence." Nor do we recall ever having written a column for "Industrial Weekly Report."

We did write a column on this subject for OGJ (March 9, p. 33) in which we referred to Venezuelan "petulancy." Could it be ... ? Grabbing our Spanish dictionary, we found that indeed it could be . . . "Petulancia, f., petulance, insolence." Now, our thesaurus lists exactly 97 synonyms for "petulance," ranging from "ascerbic" to "waspish," but never does "insolence" appear.

Oh, well, just semantics, we reckon. But what of "Industrial Weekly Report"? Could it be they were referring to the Journal as an industrial weekly report, which of course it is? Must be.

BUT reading on, we were even more startled to see what translation via dictionary can do to twist one's words.

" 'The declaration not only maintains that it would be good to have the Canadians at hand,' says Gardner, 'but it is also a warning, reminding Venezuela that for many years Canada constituted one of the best buying markets of Venezuela, yet it can play the part of the fox when pushed'."

Ouch! That ain't what we said!! We had quoted the Canadian embassy as saying it was not the "skunk at the picnic" in the U.S. imports program, and in an attempt to be funny, had added that "the statement wasn't aimed solely at proving Canadians are nice to be near." Good to have the Canadians at hand? Well, sure, but that wasn't precisely what we meant.

As for the fox...we said skunk and we meant skunk. Could it be...? Grabbing our Spanish dictionary, we found that indeed it could be—The Spanish word for skunk is "zorillo, zorrino." OK, looking for "zorillo," we first run across "zorra, f., fox," followed by "zorrera, f., foxhole," "zorillo, m., skunk," and zorro, m., male fox."

Oh well, we only hope the Venezuelans weren't offended by our calling them "insolent" in this distorted version of our copy, and as for the Canadians... well, a "zorillo" by any other name still ain't nice to be near.

The Tangerines

Alvaro Franco **June 8, 1970**

"Tomorrow," said Ingeniero Abundio Juarez Mendez, Pemex's superintendent for Mexico's Northern Zone in Tampico, "we want you to fly to our new offshore platforms. The helicopter will be ready at 10 a.m."

Turning to my good friend and guide, Fausto Munoz, he added:

"Please show him everything there is to see—including the tangerines."

I smiled. The tangerines this time of year must be very special and juicy for Juarez to be so proud of them.

I concluded my interview, thanked Ing. Juarez for his time and kindness, and spent the rest of the afternoon working on shore.

Next day at 10 on the dot, the helicopter lifted from the Tampico airport. During the flight with my guides, conversation eventually turned to food.

"We'll eat at Platform XX," one of them said. "They tell me the food is better there."

"And the tangerines are probably better," I added.

They all laughed, and I did, too.

We landed at Platform XX on schedule and had a royal welcome.

I talked to all the key men at the platform, taped lots of interviews, took scores of pictures, and got hungry.

Lunch time finally came, and with it a seven-course meal including juicy steaks and other delicacies. But, surprisingly, no tangerines.

We chatted after lunch for awhile, and then one of the engineers asked very politely, "Would you like to see the tangerines, Alvaro?"

"Sure," I replied, and remained seated, waiting for the waiter to bring them.

"Let's go," the man said, getting up.

Puzzled, I stood up and followed the crowd.

We climbed down several stairways to reach the production deck.

"There they are," he said, pointing at two monstrous spheres—two 13 1/2-ft-diameter metal survival capsules into which as many as 28 men can enter to escape fires, explosions, or other catastrophes on the rig.

They did, indeed, look exactly like tangerines in shape and color. The Mexicans had given them a most appropriate name.

I went inside one of them, along with a couple of dozen other men, the safety doors slammed, and the "tangerine" plunged into the water.

While I marveled at these wonderful survival spheres, self-contained and self-propelled, a thought came to my mind: Tom Thumb had sailed in a nut shell, Wynken, Blynken, and Nod one night sailed off in a wooden-shoe, but who ever expected the editor of Petroleo Interamericano to travel in a tangerine?

Going first class

Leo R. Aalund **July 6, 1970**

"Look mate, just put it up to the window. You don't have to look through it. Now tilt it so you'll get just a bit of the engine pod. Now snap it."

Without looking back, this editor did just that because he was losing the struggle for headroom with the Yashica in the passenger lounge. The Pan American 707 was climbing rapidly, and the Rio beaches of Copacabana and Ipanema were fast fading under the wing.

Turning around, I looked across at the passenger with the peremptory photography instructions who had joined me in the lounge. He was British, big, and very angry. His companion was also British and angry.

Introductions seemed in order, at least to find out what the trouble was. Then their story flowed forth. They had come to the lounge from the tourist section to speak with "the captain of this aircraft," who was a few feet forward in the cockpit guiding the jet toward Miami. They were a tired cameraman and technician for the BBC (indeed experts) going to the states for their first 2 days off in a month. They had spent a month in the muck of earthquake-shattered Peru. Then it was 2 sleepless nights in a Rio street waiting for the return of the kidnapped German ambassador.

Their tickets said first class, but at the Rio airport before takeoff they were told they were going tourist or not at all. When tempers flared, a little airline girl, they said, told them to go fly another airline. That was too much.

"Blokes like us who put up with grime and danger in our job daily have to relax in style sometime or fall apart."

Soon the smiling American captain appeared. For openers the blokes dropped "BBC" on him. The sympathetic officer began taking extensive note of the affair. Shortly, the purser and two stewardesses appeared. It was cocktails, snacks, white table cloth, red carnations, canapés, lobster, wine, roast duck—in the lounge—accompanied with big smiles. They just couldn't do enough.

Typical service on this airline? Perhaps, but this OGJ bloke wasn't about to go back to his aisle seat to find out. Finally, the captain left, with assurances that letters would be written. The television blokes weren't convinced though.

"I can hear him now, Roger, up there saying, 'We got a couple of British goons on board making trouble'."

Later things got sleepy over the vast Amazon Basin. Over Haiti, the Britons relaxed a bit and asked me for hotel recommendations in Miami.

"Why don't you spend a couple of days at Nassau in the Bahamas. It's only a half hour or so from Miami." That was the perfect suggestion. Goodbyes were said in Miami.

Late that night this groggy editor was over the Gulf of Mexico flying at midnight to Houston on another major American airline.

A pretty stewardess, who had all of 10 passengers to serve, asked in a commanding voice, "Cocktail, coffee, or milk, sir?"

"I would like a glass of iced tea, please."

"Iced teee!" The reaction was bitter lemon. "Well, I guess I can make some."

One can imagine the galley talk, "Dora, there's a goon back in economy that wants iced tea. Some people sure have gall."

And my two blokes, when I needed them most, were by now sleeping in a hotel (first class) in the Bahamas while the manager undoubtedly was roping off a private section of beach for their morning swim.

Clear running water

John C. McCaslin **July 27, 1970**

Did you ever visit a "foreign" city that was more U.S. than U.S.? Well, there's a big and fast-growing one up where the Bow and Elbow rivers meet just east of the Canadian Rockies in Alberta.

This story began when Col. Macleod set up a Northwest Mounted Police camp in 1875 to watch the Indians, flush out the cattle rustlers and whiskey traders, and bring a bit of calm to a wild and lawless land. The Mounties decided that this junction of clear running water should be tagged Calgary, which in Gaelic means just that. And, to this day, Calgary keeps the running water clear, despite a trickle of pollution from up Banff way.

The trading people came into Calgary after 1875. By 1883, the town had a "whopping" population of 600 lusty souls. The Canadian Pacific Railroad reached the town that year; by 1890 there were more than 2,000 people in the dusty valley. Fast growth continued in the wake of the steam engine. Lush grazing lands lured cattle ranchers from the south. And, like a chapter in any good Western novel, the cowboy appeared on the scene. Free homesteaders followed from the East, and Calgary became a chartered city in 1893.

From 1893 into the new century Calgary bustled and grew. Prolific wheat fields mingled with green cattle ranches to set the stage for one of North America's most beautiful and prosperous cities. For years flour elevators and the Palliser Hotel dominated the skyline along Ninth Avenue Southwest, hugging the railroad tracks. By then the railroad had pushed on over the majestic Canadian Rockies to the Pacific Ocean.

Then it happened. From not too many miles southwest of "Clear Running Water" came the magic word one day in 1914. You guessed it. "Oil." Turner Valley became a household and ranch-land word. But, oddly enough, the excitement died down; and Calgarians went back to the meat and flour life that built their city. The 1920's and 1930's came and went. Oklahoma City, Burbank, and East Texas were so far away indeed.

Then, one day in 1947, Canada joined the family of oil nations in one fell swoop at Leduc, just south of Edmonton. Oh, yes, Edmonton is that friendly rival city to the north of Calgary. Edmonton was already larger than many Canadian cities when oil changed its history. Suddenly everyone forgot that Edmonton was capital of the province, that wheat and meat were Alberta's lifeblood, and that Oklahoma City, Burbank, and East Texas were so far away.

The rest of the story is one of prolific growth for both cities and the entire province of Alberta as some of the world's largest oil and gas fields spewed new blood into Canada's vast western domain. Enter the oilman, his gear, and the accompanying industry and wealth.

But during the week of June 19-26 this year came the largest single group of people in such a short time into Calgary. By camper, car, truck, plane, and train the geologists of AAPG flocked. The occasion was the largest convention yet for Calgary and nearly the biggest yet for AAPG at more than 4,300 registered rockhounds. And what better time to have a meeting than June near the foot of the rugged Rocky Mountains. In fact, for a couple of weeks after the meeting it was hard to find a geologist in his office back in the States. They all took off out of Calgary, venturing up the Clear Running Water to Banff and Lake Louise and other alpine gems to discuss pollution, Strachan-Ricinus, Rundle limestone, and the Devonian D3.

There's one sour note here. To quote a local Calgary barkeep. "Vas ist ein geologist?" Of course, this lad had just come to Canada from Munchen.

Feeling the pulse of Alaska

Howard M. Wilson **August 10, 1970**

Your plane edges the southern coast of Alaska en route to Anchorage; you're awed at the magnificent white Chugach Mountains, their massive glaciers oozing out into the Gulf of Alaska.

Your pilot cuts the altitude as you near Anchorage, and you skirt the tops of more mountains, snow-streaked and showing signs of summer thaw. As you approach a landing, the green valleys of spruce, birch, and aspen rush into view.

This is not Los Angeles, or New Orleans, where the sense of adventure has long since worn thin. This is Great Land. It's exciting to be here.

Your fellow passengers sense it, too. As they ride the limousine from the airport to the Anchorage Westward and the Captain Cook, they talk to each other. Strangers, yet.

Two elderly and slightly daring women travelers from Dallas—tourists, not oil wives—are looking for a "different" vacation this year. Three workmen just in from Attu are looking for just plain fun. The young high-school graduate from Longview is looking for work. And the Journal's West Coast editor is looking for copy.

The two women have no hotel reservations. The driver warns them: "You made a mistake." They are unworried, even giggly: "We can always sleep in the park."

The driver, a fount of information, also warns the young man from Longview: "There are no jobs. I worked on the North Slope last summer building a road—and look at me now."

The Texan is not disturbed: "I can always find a job."

The men from Attu have been to Anchorage before, but the weeks of privation in their western outpost have whetted their appetites for food, drink, and recreation. "Stella should be back from Seattle by now," remarks one. The others cackle.

It's late evening, but the sun is still high in the west.

"It never gets completely dark this time of year," explains the driver.

The limousine passes through a drab section, and you see a cafe called the "Garden of Eatin'." What happened to the Malemute Saloon?

Anchorage, you find, is no Paris of the North. But there are new buildings downtown, and others are rising. You're surprised at the long-haired and bearded youth on the street. The hippies and semihippies have heard about the unspoiled country of the North, and they have driven the Alcan highway in their old cars and campers to make the scene.

As if by design, "Easy Rider" is showing on the main drag.

The Baptists are strong here, too. The billboard says there are 20 Baptist churches (Southern, no less) in the Anchorage vicinity.

And the children. They are on the streets everywhere. The limousine driver has the statistics: "The average age in Alaska is 27, and 60% of the population is kids."

And the tourists. Like the two ladies from Dallas, vacationers have discovered the Great Land. The hotel lobby, you find, is awash in suitcases, their sightseeing owners checking in or checking out. Business of the Gray Line Tours is thriving.

And, finally, the oil crowd. A few days of visiting in Anchorage convinces you they like it here and want to stay. If oil activity isn't flourishing at the moment, they tell you to wait a year: "When we get that pipeline going . . ."

A visit out west!

Larry Auldridge **September 14, 1970**

Dear Boss:

I've been trying to get around to tell you how much I enjoyed my recent visit to Tulsa, but you know how hectic things can get in New York City.

Anyway, my first visit to Tulsa was quite pleasant, quite interesting, and, I might add, quite quiet. You weren't having a holiday there, were you? Or maybe a strike?

Probably it was a strike. You know, the kind we're always having—subways, buses, trucks, sanitation men, delivery men, post office. Even the Rockettes at Radio City.

But everyone was friendly. The waitresses at the hotel—obviously not on strike—were very nice every morning.

But when I walked from the hotel down there to the Journal office, I noticed that traffic was light. Must have been a pleasure for those automobile drivers not to have to fight the traffic. The pedestrian street signs were working, and even some of the people waited for them to turn green. They smiled at me—very politely—when I just went on across before the next bunch of cars came along.

And, I must say, the air and sun were nice. As I say, you must have been having a strike. Building maintenance men, maybe? Or maybe factories? There wasn't a cloud in the sky—God-made or man-made.

It was the same when we went to lunch. Just nice fresh air. I guess the people who weren't out on strike were enjoying it, though. They even spoke to each other. And even if they didn't know us, they sort of stepped to the side to let us walk by.

I guess business was pretty slow, too. We were waited on right away. And even back at the hotel that night, one of the hotel bellmen saw me take out a cigarette and came over with a cigarette lighter to light it. He was probably new there.

The cab business didn't seem any better, because the driver opened the door for me. And when he reached your house, he actually said "Thank you," when I tipped him.

I must say, too, the evening at your house was very pleasant. Your neighbors must not have been home, either. Because I didn't hear a peep out of them. It was like being in the country. You know, way out West—if I may use that phrase.

Incidentally, I meant to mention it then, but forgot. But were you having some sort of electric power "brown out" like we have here? The night before, I took a walk around the hotel and I just didn't see any—or many—neon lights like the ones they have for night clubs and restaurants.

There must have been some sort of curfew, too. Because there weren't any stores open at night. I looked for a drugstore, but they were all closed.

And maybe you were having a police "sick-out," as they call it here. You know, police can't go on strike, but they call in sick. I saw a police car, but I suppose the others were in the garage because of the "sick-out."

Anyway, I went back to my room early. About midnight, I guess. The air-conditioning was on, so I didn't have to open the window. Just before I turned in, I left a wake-up call. The girl who took the call said "Thank you" and "Goodnight."

I must say, I slept like a log. And in the morning the wake-up call came. It must have been the same girl. She said, "Good morning, sir" and gave me the time. I started to hang up, but she was saying, "Have a nice day."

I looked out the window—the 14th floor—and saw the bright, clean streets and buildings. It was still Tulsa, Okla., all right, it couldn't be New York.

An editorial first

Max L. Batchelder **September 28, 1970**

Dozens of full-time editors work hard to put out the Oil & Gas Journal every week, and most of them are invisible. Invisible as far as the pages of the Journal are concerned, that is.

There are a couple of exceptions, of course. John McCaslin is shown on page 32 of this issue, for instance. And the pictures of both Frank Gardner and Gene Kinney accompany their columns each week, but that's about it. Like most publications, the Journal has a thing about its own people muscling into pictures that are to run in the magazine.

It's a good thing, too. Frank Gardner and Gene Kinney are the only two people on the staff who look like editors. The others look like plumbers or lab assistants or bridge-painters. Before I got into the publishing game, I always figured editors were thoughtful-looking fellows smoking thoughtful-looking pipes.

When I came to the Journal 13 years ago I was stunned to find no one smoked a pipe except one man who turned out not to be an editor at all. As a matter of fact, for 2 days I thought I had been assigned an office in the middle of the printing department until I found that all those harried-looking, fast-paced folks were editors. I wondered how in the world such a crew could put out a weekly oil magazine.

It turned out that this unlikely-looking bunch knew more about petroleum and how to write about it than any other collection of people in or out of the country.

A roomful of Journal editors can answer almost any question about the petroleum industry. Their answers will be knowledgeable and current. Collectively, they are like a huge computer programmed to supply anything anyone might want to know about petroleum.

This tremendous backlog of oil know-how shows up all through the pages of the Journal—but only an editor's know-how appears, never his face.

Journal editors just don't show up in Journal-run photos. Oh, there was a crowd scene at the International Petroleum Exposition that included a Journal editor—all you had to do was pick him out from the 5,000 other people shown in the photo.

And one time a Journal editor almost made the cover—his hands appeared in a photograph, shuffling around a lot of statistical material.

But I've decided to break that taboo once and for all. Quietly, without alerting management, I'm going to show you readers a picture of our pipeline editor, John O'Donnell. That's him there in the helicopter, taking pictures of a pipeline. I'll bet you had no idea he looked like that, did you?

Now that the taboo is broken, I'm sending along a picture of myself to Editor George Weber, hoping he'll take the hint. You wouldn't believe what I look like!

Getting to the top

Robert C. Ewing **October 5, 1970**

Even back in our engineering and construction years we got into a lot of refineries. And rarely did we decline an invitation to climb or ride to the top of the tallest structure in the plant.

In small plants, the trip was a hard climb, because usually the only platform was at the top of a tower. Rest stops were made leaning against the ladder cage. More prosperous plants had platforms at every manway and climbing was easy.

Another way to the top is via a headache ball suspended from a crane boom. One does not decline this invitation for fear of snickers and guffaws questioning one's virility.

This mode of travel provides an inestimable charge to our ego, which has a slight edge over our fright. An added thrill is the mechanical failure of the lift while in midair, whether real or pretended.

Rewards received at the top of a refinery structure vary.

Sometimes motion sickness threatens us on a tall thin tower under gusty wind conditions. On a structure downwind of a heater stack, we encounter hot and oxygen-deficient air. This situation compares to one of the alternatives in the hereafter.

Lately, plant public-relations people have learned to operate the elevators on cat crackers and coking units. We can't let a PR man top us, so we take a slow ride in a cage bigger than a bread box but somewhat smaller than a phone booth.

On the way up, we are pelted with loose catalyst or hot bits of coke, depending on the structure, through openings in the cake. Flashing lights, odd noises, and the wretched side motion of the elevator is about as weird as a drug-induced trip.

Our latest trip was to the top of an unusually tall cat cracker. The reason was to get a better feel for the results of the latest pollution abatement devices installed in the plant. We were promised no heat or odor, and the elevator sounded like fun.

Once at the top of the unit, we looked to the south to view this fine refinery. Views to the east and west equaled the view to the south—no smoke, no odor, and we breathed well.

However, our glance to the north occurred simultaneously with a loud clap of thunder—real thunder and rain. A fast moving front had swept in on our blind side and surprised one of the aforementioned alternatives out of us. So with our host, we scampered back up three flights of stairs in a now driving rain to the elevator door.

Slotted platforms and stairways offer no shelter from rain, and elevators are a poor second choice. They leak too. Upon reaching the ground we raced to the control room with wet clothing plastered to our skins. Our ties—well, there is nothing more ridiculous than a water-logged tie in a refinery.

Our record stands. We have achieved a long line of blemishes from making trips to the tops of tall towers. We always manage to tear a jacket, cut a shoe, get dirty, burned, suffocated, or we drop our camera. This time, we got drenched.

You might think we would stop this silly stuff after 25 years of flirting with disaster. Our better senses always say, "cut the comedy." But our egos talk with louder voices. When we are in a plant and the invitation comes, we put our hard hats over our egos and go all the way to the top.

And now, a message to our stringers ...

Dr. Frank J.Gardner **November 9, 1970**

"How can you possibly get to all those places? You must be on the road constantly!"

We've been asked that question so many times, we finally settled on a canned answer: "I don't, and I'm not!"

Well then, how do the international editors of the Journal manage to cover the news in the hundred or so different oil-oriented countries of the world? By oil-oriented, we mean they either have oil production, refining, exploration, or pipeline activities (we daren't try to include service stations!). Some have all these, some only one or two, but if there's the slightest oil activity there, we're supposed to cover it. Our beat takes in the entire world outside the U.S. and Canada.

We have a lot of help—the time-honored"stringer"or correspondent—and often he doesn't get the credit he deserves. All over the world, we have these part-time reporters who, once a week, send us the oil news from wherever they are.

Each is supposed to cover only his/her own territory, and get paid only for what actually gets into print—a fixed fee per printed column inch.

But sometimes they may overlap. Our man in Pakistan may yield to temptation and file a story on India. That's a no-no, but if our man in New Delhi has failed to file the same dope by presstime, we accept it from the Karachi stringer. If both of them file the same story, the credit goes to the man in India.

You can see how many problems might arise, and sometimes we have to make like Solomon and split the credit. But none of our stringers has ever complained.

They're a fine lot, too. Some are professional reporters, some are moonlighters with oil companies, some are teachers, engineers, or government workers. We even have one housewife who does a bang-up job—and one man who works for an airline!

What's interesting is the way their personalities often shine through their correspondence. Some are very proper, with never an aside to the editor. But others attach humorous remarks and personal comments that add spice to the job.

Particularly adept at this is our man in Israel. Sitting on a powder keg as he does, a sense of humor (he says) helps a lot. He salts his copy with such expressions as "'Robbers!' cried the pipeline company when tanker rates hit the ceiling.

Or how about this: "The impression one has out here is that Middle East producers are slow to realize that the good old days are gone for good, and one has to live and produce in this new world, in which logic doesn't count, and persuasion and promises given one day aren't worth the paper they're printed on 1 hr later. Still, it's fun to be around here—as long as you don't get blown up or hijacked, that is!"

Then there's our Russian correspondent, who takes little vacation trips and sends us postcards snidely scrawled "Thinking of you!"

But wherever they are, and whoever they are, we couldn't work without 'em. They get little enough credit, so we'll just take this opportunity to say out loud—"Thanks, fellers!"

How big Indonesia?

W. A. (Bill) Bachman **November 30, 1970**

The weirdest oil story of 1970 has to be an article appearing in the Zurich Die Weltwoche under the byline of Joachim Joesten. Thanks to Journal reader Henry F. Lippitt II, Los Angeles, we got a copy of the article, which had a real frightening political consequence in this country.

May we quote Herr Joesten:

"American scientists have come upon a lake of oil 4,000 miles long and 300 miles wide, stretching under the surface of the sea from the Gulf of Siam across Indonesia all the way down to the northern Australian coast.

"The offshore deposits alone in this region are estimated by experts as totaling three times all other offshore deposits in the world."

A little farther, Mr. Joesten quotes an American geologist as saying: "In 5 years the offshore fields of Thailand, Cambodia, Malaysia, South Viet Nam, and Indonesia will be ready to produce 400 million b/d—or more than is now produced in the entire Western world."

And as a clincher, Mr. Joesten indicates that the entire American military effort in Indochina has as its prime objective providing safe and secure onshore support to the offshore drilling campaign.

Granted, oilmen are surprised at nothing. But come now! An oil pool 4,000 miles long and 300 miles wide? And production of 400 million b/d? The reporter must have misplaced a decimal point by two places. You'll pardon us if we doubt these figures.

The disturbing aspect of this article, however, is that it was picked up by a candidate for Congress in Utah and used to assault both the oil industry and the Nixon administration.

A. H. (Bob) Nance, a Utah Democratic candidate for Congress, in an article in the Salt Lake Tribune, attempted to give the Die Weltwoche report substance by saying the Oil & Gas Journal had carried brief stories confirming U.S. oil involvements in Southeast Asia.

The Journal, for the record, has carried stories on discoveries in Indonesia, on new drilling, and on leasing interests of American companies.

But that's all. Nowhere have Journal stories indicated or predicted reserves so massive they would support production of 400 million b/d. Such statements are hogwash.

Mr. Nance, though, compounded the slurs tossed out by the Die Weltwoche article with a few low-punch questions. "Has the U.S. been involved in the Viet Nam war to protect human rights or drilling rights?" he asked. "Has the U.S. paid a price of over 50,000 war dead and hundreds of thousands of casualties to protect the oil concessions granted to these U.S. oil companies?"

The fact is, America's involvement in Viet Nam occurred long before the current oil play in that part of the world heated up. Mr. Nance impugns the ethics and patriotism of a host of people, including Democratic Presidents Kennedy and Johnson and Republican President Nixon plus hundreds of leaders in the oil industry.

This just has to be the weirdest oil story of 1970. Apparently the people in Utah's second district agree, since they did not vote Mr. Nance into office.

Go ahead. Breathe.

Lester Van Dyke **March 1, 1971**

We heard some encouraging words here in New York City recently. That's right. Encouraging.

We may now breathe deeply 87 days out of the year. This may not sound encouraging to someone living in Arizona. But considering that we were only "allowed" to breathe deeply 38 days in 1969, it's a real improvement.

It was, however, still recommended that we hold our breath on 80 days last year—but this too is much better than the 114 days of anoxia we suffered in 1969.

Days in which we really shouldn't be breathing, but could at our own risk, declined from 209 in 1969 to 194 in 1970.

It occurs to us that all this may account for the great rushing about often commented on by visitors to the city. Since the average human can only hold his breath for say 1 or 2 min, New Yorkers really have to hustle between buildings to avoid inhaling too much.

But back to the story. The first few paragraphs are obviously a very loose translation of some truly encouraging news just released by the New York City Department of Air Resources. Incidentally, the words they actually use to describe air quality in the city and metropolitan region are "acceptable, unhealthy, and unsatisfactory."

The outlook is that we may be able to take deep breaths on even more days this year. Obviously there are a number of pollutants involved in the air quality measurements. But the Air Resources Department (ARD) is clear in stating that the main reason for the improvement in air quality last year was the reduction in the sulfur content of residual fuel and bituminous coal. And mainly the word reduction applies to resid, since it and other fuels were replacing coal.

Sulfur content in fuels has fallen from well over 2.5% in the mid-1960's to the 1% and lower regulations now required in the metro region.

On a more subjective level, I'm enjoying my mile walk to and from the bus terminal and office each day a lot more this winter than last, because my nose and throat don't burn nearly so much or often as they did the first 3 years I was here—my "transplant" from Tulsa came in January 1968.

There have been other benefits, too. Blending of high-sulfur Venezuelan oils with low-sulfur North African crude also cut particulate emissions by "factors of 2 to 4," the department notes.

In one sense that may not be as good as it sounds, though. Used to be a man coming home from the office with dark smudges on his face could mutter something about "darn fly ash," quickly putting an end to any question that it might be his secretary's mascara. But take heart. There's still probably enough in the air for an alibi.

It's clear the efforts of concerned citizens, the politicians, and the willingness of the oil industry to cooperate are paying off. But "obviously we still have some way to go to reach our air quality standards for sulfur dioxide in the city, which is .03 ppm as an annual average," the ARD says. This points up the importance of the provisions in the air-pollution-control code now before New York City, and other cities, which will require even stricter sulfur limits in fuels of the future, it continues.

There are costs the consumers will have to pay for the lower-sulfur fuels, to be sure. And sensible, stationary targets must be set up for industry to meet. But we shouldn't take too much pride in these figures, not until we can take deep breaths 365 days out of the year.

Crossing the Orinoco

J. P. O'Donnell **March 8, 1971**

Al Franco, editor of our companion publication for Latin American oil people, Petroleo Interamericano, has written an article for The Journal that brought back some old but still rather vivid moments.

Al wrote about the installation of a pipeline crossing of the Orinoco River in Venezuela. He described the river as "one of the wildest and longest in the world." I can't vouch for either of these superlatives, but for a few moments at one point in my life it was the scariest.

It was an incident that happened many years ago, long before the Orinoco was spanned by a bridge from Ciudad Bolivar to the then tiny village of Soledad. In those days a ferry ran back and forth across the river which, as I recall, was about a mile or a mile and a half wide at that point.

I was with a couple of fellows in Ciudad Bolivar, and one suggested we take the ferry across the river for a look at Soledad. We did; the look was very brief, for you could get all around Soledad then in about 5 min. With some time to kill, we went to a sort of bar-restaurant-hotel combination for a beer.

We were well along on what might have been the third beer, when one of the local citizenry stuck his head in the door and told us, "If you are going back to Ciudad Bolivar today, you'd better hurry. The last ferry will soon pull out."

We finished the beer in a few gulps, paid the check, and ran down to what served as the ferry slip. Before we got to it, we saw that the ferry had already pulled out. One of my companions, more fluent in Spanish than the rest, said, "I'll see if I can't get one of these boat owners to take us across." There were a few skiff-like boats equipped with outboards tied up nearby.

It seemed to take him quite a bit of time to make the deal, but he did. Then he turned to us and said, "Let's go. We've got to hurry."

We were only a short distance offshore when the wind began to pick up, and the river became choppy.

The waves were, so I remember, about 2 ft high when the motor sputtered and died.

The boat tossed about while the owner tried to restart the motor. Time after time he'd make an adjustment, pull the starter cord and nothing would happen. It didn't take long for us to realize that, even if the wind didn't get any stronger, there was a good possibility that the boat would capsize.

With that realization came thoughts of things I'd read about South American rivers. I remembered seeing signs cautioning people to stay out of and keep their animals away from streams infested with bilharzia. Somewhere I had read or heard that the Orinoco teemed with piranha and electric eels.

I resolved then that if the boat did capsize, I would not try to swim to shore but would try to crawl up on the bottom of the boat and stay there until it drifted ashore or we were rescued.

Two things happened that were not at all reassuring. One of my companions found some life preservers and donned one. As he pulled it tight, it came apart.

The other was even less reassuring. It appeared to me that people were gathering on both shores. It didn't occur to me that they may not have had anything that they could use to try rescue. All I was aware of was that no one was coming.

Fright can breed some awful thoughts, and my mind was full of them. One that never occurred to me was the thought of drowning. The alternatives were so much more horrible than mere drowning.

Needless to say, nothing disastrous did happen. If it had, you wouldn't, of course, be reading this. I can't say and don't remember how long it was before the motor did get going. It seemed like agonizing hours longer than the few minutes it must have actually taken.

If Al writes a story for the Journal about Colombia, that'll furnish me an excuse to do another Journally Speaking. There's a still better story to tell about getting lost in the fog-shrouded Andes in a single-engine plane.

The greatest mystery

Henry D. Ralph **June 28, 1971**

While preparing the story on gasoline prices this week, OGJ resurrected this still timely column written July 8, 1963, by the Journal's retired Chief Editorial Writer, Henry D. Ralph. We thought it worth reprinting. So here is "The Incident of the Increment, or What Became of the Barrels?"

Once upon a time there was a Smart Refinery Operator. He was an Engineer, so he knew all about Figures and how to calculate Costs.

His refinery could produce 100,000 b/d of gasoline. But the Company was selling only 85,000 b/d. So his output was cut back to that figure.

The Company wasn't making as much money as it wanted to. So all departments were told to cut costs. The Refinery Operator went to work with his slide rule.

He figured the price of the crude oil, the labor, and the various overhead items, and came up with a figure on what it cost him to produce a barrel of gasoline. He couldn't see any way to make it cheaper so long as he was held to 85,000 b/d.

But that 15,000 b/d of unused capacity worried him. So he did some more figuring.

Then he made a Great Discovery.

He discovered that he could run his plant full blast without any additional cost in labor, investment, or overhead. Those items were already being charged against the 85,000 b/d he was producing. So he could produce an additional 15,000 b/d at no Cost whatever except for the crude.

He even invented a name for these 15,000 bbl. He called them Incremental Barrels, because they were an increase over what he would be producing otherwise.

Then he went to his Top Management, and said:

"Look, Boss. I can give you 15,000 b/d of gasoline at 4 cents a gallon under the cost of the 85,000 b/d you are now getting. That will increase the Company's profits. How does

that sound?"

"Wonderful," said the Boss. "Hurry up and do it."

The Boss sent a memo to his Marketing Man, saying:

"Sell an additional 15,000 b/d of gasoline. We are getting this at 4 cents under the market price, so anything you get for it will be gravy."

The Marketing Man couldn't get rid of the Incremental Barrels through his regular channels. And he couldn't find anybody who would pay the going price for them, because everybody seemed to have enough gasoline.

Then an Independent offered to buy the Incremental Barrels for 3 cents a gallon under the usual price. Since the Company would still make a penny a gallon, this seemed like a Good Deal.

The Independent sold the Incremental Barrels to motorists at 3 cents under the retail price. This started a Price War. All the other stations reduced their retail prices, too.

At the end of the year the Boss discovered that the Company had made less money selling 100,000 b/d than when it sold only 85,000. He called the Marketing Man on the carpet and demanded:

"How come?"

"Price Wars," said the M.M. "They pulled down prices all over."

"Who starts Price Wars?" roared the Boss.

"The blankety-blank Independents," said the M.M. "We have to follow their prices down. It's murder."

"Can Independents operate stations more economically than we do?" inquired the Boss.

"No indeed," bristled the M.M. "We are a Model of Efficiency. We watch all Costs."

"Then how can Independents afford to sell so cheap?"

"They must get gasoline for less than our stations do."

"Where do you suppose they get low-cost gasoline?" mused the Boss.

"Search me," said the Marketing Man. "That's the greatest mystery in the oil business."

Lilacs in the Kremlin?

George Weber **July 12, 1971**

The eighth World Petroleum Congress will go into the record books as a success. But of a different sort from that intended.

As a purely scientific and technological meeting of minds it fell short of the mark, according to most delegates we interviewed during and after the Moscow conclave.

The papers ranged as widely in quality as in subject matter. A few were exciting. More were reasonably enlightening, and many were dull, or dreadful, or both.

This stems from the lack of discipline inherent in a multinational organization. Each national committee decides what it will bring to the program and standards of quality varied.

Some of WPC's problems are built in. Red tape and ponderous procedures make the liveliest of papers dull when read 8 months after writing. Those authors did best who merely sketched their basic papers and spent their short valuable podium time in updating, commenting, and answering questions.

But there again something was missing. Soviet host Shashin put his finger on it when he scolded the delegates in the closing ceremony for following their wives off on junkets about Moscow, and for failing to do their homework by ordering and studying preprints in advance of the meeting.

He might have included some session chairmen who allowed time and crackpot comments to get out of hand to the detriment of good solid exchange of information.

So the Congress failed to attain its high purpose. But it was still a rousing success as a meeting of oilmen, though this may displease the purists who run the show.

What catalyzed the whole affair was the massive exposure for the first time of some 800 Russian oil men plus several hundred others from Eastern Europe to some 2,700 oilmen from countries outside the Soviet sphere. And that latter group just happened to include a generous sprinkling of those who serve the international oil industry—the manufacturers, licensors, suppliers, and service-company representatives who have a large hand in making petroleum a viable energy business.

Here, brought together in mutual interest on an unprecedented scale, were the "know-how" of the western industry's technologists and the "want-to" of the Soviet oilmen from field, refinery, and petrochemical plant. And linguistic barriers aside, they really talked the same language: how to get the job done.

The Journal is just Philistine enough to applaud. Some business was done and a great deal more will probably evolve from a mutual understanding and admiration that binds the doers among oilmen whatever their political or geographical backgrounds.

All of which makes a pretty good point on how this international oil business manages to serve the entire world with most of its energy. With all due credit to the thinkers and planners, oil is not found, produced, pipelined, or processed altogether by calculus. It takes the operators to translate theory into barrels per day or tons per year. And fortunately, there were enough of those "heroes of the industry" present to make a go of the meeting.

We may not buy one another's ideology, not by a long shot. This meeting brought no political conversions one way or the other. There wasn't time. But what it did serve to do was to transcend both business and science in offsetting to some extent the dehumanizing regard of opposing peoples who never get to say hello to one another.

A lot of our preconceived notions went down the drain. Maybe some of theirs did. One of our surprises, which is no doubt irrelevant, was the sight inside the grim walls of the Kremlin of lilacs blooming profusely in the waning weeks of a mild Moscow spring.

Westward ho!

John P. O'Donnell **August 2, 1971**

The mood of the times is readily apparent to a visitor to the Middle East these days. You are frisked thoroughly and your hand luggage searched equally thoroughly, even if you are only an intransit passenger and get off at an airport just to break the tedium of travel.

It is a bit over 24 hr from Houston to Tehran via Chicago, London, and Tel Aviv. That's a long ride. So, when you get a chance to get off the plane, you take it. I did so at Tel Aviv where the inspector was so thorough he punched each key on my portable typewriter. He must have thought that one might be connected to something, possibly an explosive. And, you know, before he had finished, I did too.

There were armed guards at the Tel Aviv, Tehran, Damascus, Beirut, and Istanbul airports. At one airport, passengers were warned not to take pictures. At another, in-transit passengers were told to remain aboard. A third had anti-aircraft gun emplacements along its runways.

Even though there was a cease-fire between the Arabs and Israel at the time, you had a sense of tension. It was probably more mental than real, for the sight of armed men with a purposeful air about them is disquieting.

One of the precautions had an interesting effect. Iraq won't allow commercial planes to fly over its territory. Therefore, to go from Tehran to Damascus, you fly west through Turkey, south over Cyprus and then east over

Lebanon to Syria. In eastern Turkey you pass close enough to Mt. Ararat to get a good look at it.

Its snow-capped peak towers 16,946 ft, and it's the highest point, I believe, between the Alps on the west and the Alborz Mountains on the east. If you ever wondered why Noah used it as a base of operations, this should answer your question.

One day in Iran was spent on a tour of the Agha Jari and Marun fields and the Bid Boland gas plant. The fields are in the foothills of the Zagros Mountains.

These hills are crossed, crisscrossed, and recrisscrossed innumerable times by goat paths. You get the impression the goats knew they could not use one path indefinitely and, therefore, would start a new one. You also got the impression that some of the paths were hundreds, maybe thousands of years old.

Pipelines also crisscross the area, gathering oil to be shipped south to Kharg Island or gas to be shipped north to Russia. They must have seriously disrupted the travel pattern of the goats. But the disruption must not have disturbed the goats themselves, for they obviously are not neurotic and the herds are numerous and large.

Conservationists who have so effectively blocked the Alaskan pipeline would be horrified at what went on in Iran. The possible discomfiture of a few thousand caribou in Alaska has them terribly upset. The dislocation of many thousands of goats would have shocked them to the core.

Building pipelines did help Iran's economy. It can now afford to irrigate formerly unproductive land. It has roads, schools, and hospitals that didn't exist before oil. Tehran's streets teem with traffic, and getting a cab in the rush hours is as hard as it is in London or New York.

They call Esfahan the Pearl of Persia. Its Shah Abbas Hotel is one of the world's most spectacular hostelries. It has fine eating places where you can get such tempting treats as chelow fesenjan, chicken cooked in pomegranate sauce, and the world's best caviar, a bit on the gray side and not at all salty.

However, conservationists might argue that better living for Iran's citizens and her growing tourist attractions are surely not worth bothering the goats. It would have been far, far better, according to their thinking, that the goats be allowed to pasture where they please and that the Iranians trade in their Volvos for camels.

Arcticana — Part 2

John P. O'Donnell **November 1, 1971**

Midsummer is really not the best season for a first-time visit to the Arctic. True, you see it wearing its most attractive face, but you don't see the Arctic as it is in winter. And there's a lot more winter than summer north of the circle.

Much of the Arctic looks green and rather friendly in early August. That appearance changes abruptly when you fly over the Romanzof Mountains in northeastern Alaska. They are a bunch of knife-edge rocks that stick up in the air—like 9,239-ft Mt. Michelson. Intervening ravines are filled with snow to no-one-knows what depths.

The expanse of greenness is misleading, for you know very well it is not what it seems to be. You know that under the fragile and vulnerable green tundra the permafrost covers a vast area. At times you can see expanses, many miles in extent, of ice-formed polygons. Every once in a while you'll see a pingo, a thrusting ice mound that rises out of the flatness.

There is continuous and discontinuous permafrost and it's found both in the Arctic and subarctic. Incidentally, don't use the Arctic Circle to separate the Arctic from the subarctic. The Arctic, in reality, is "that area in which the average mean daily temperature of the warmest month of the year does not exceed 50 degrees F."

Our trip, courtesy of Gas Arctic, took us to Yellowknife, Fort Norman, Inuvik, and Prudhoe Bay. A flight over the Arctic Islands with a stop at Resolute on Cornwallis Island had to be abandoned because of zero-zero conditions: zero ceiling and zero visibility.

Things we did see convinced us of one thing. Those who speak of the Arctic as a challenge are voicing the understatement of the century.

A road must be built through the permafrost area for stringing pipe and other services incidental to pipeline construction. An engineer with one of the Arctic pipeline projects says it will cost $200,000/mile to build a gravel road. Another says the delicate area an Alaska-Canada line would have to cross could hit 700 miles. Simple extrapolation puts the cost of the road alone at $140 million.

Road or pipeline construction in the Arctic is viewed by "Lower-48" conservationists as a threat to the migratory habits of the caribou. An in-depth study of the caribou is being made by one pipeline group. It should provide more factual information on the "palmate-antlered deer" than any previous source.

Scientists studying the migratory habits of the Porcupine herd, largest in western Canada, have found that "the caribou often favor the seismic lines and winter roads cut by oil industry where they found the walking easier." It appears the caribou is as creature-comfort-conscious as a human.

Last week we mentioned two Arctic sports events but missed a third. It's the Midnight Sun Golf tournament, played every June on the sand fairways, oil-surfaced greens of the Yellowknife Golf Club. It has two unusual hazards which often result in lost-ball penalties.

The course abounds with granite out-crops from which golf balls ricochet like bullets. Then there are the ravens—known as Tuluga, the black bandit—

which swoop down and make off with the rolling balls.

You would think, I'm sure, that Yellowknife, with a mean daily average temperature of 57 degrees F. in August, would have no need for air-conditioning. Yellowknife has 20 hours of sunlight per day in summer, including 8 hr of bright sunshine. If you happen to get a room on the sunny side of a motel, it can be real hot by bed time.

Before concluding this two-part collation of Arctic bits and pieces, there is one more thing we found impressive. Three groups are spending $30 million on studying ways to build and operate Arctic pipelines without damaging the environment. They have learned how to build them without endangering the ecology and it's a dollars-to-doughnuts bet they'll learn how to operate them as well. They are footing quite a bill to prove that they are good citizens, a fact which those good citizens who are opposing them with less factual arguments should be inclined to recognize.

A wild ride for news

Carlos Byars **November 8, 1971**

We were 1,500 ft up and 110 miles out to sea when the pilot pointed out water-spouts, leaping from the sea like a cluster of angry snakes. He needn't have bothered. I had seen them already, and my eyes were bugging out like a voyeur at a nudist convention.

We—meaning the pilot, PR man, cameraman, radio repairman, cook's helper, and myself—already were making our second effort to reach the Dixilyn 9 jackup, drilling on Pennzoil Offshore Gas Operators' lease in West Cameron Block 587.

Having dodged the iron on rig floors all over West Texas, I was curious about what life is like offshore, where there is no place to run and the nearest supply store is a hundred or so miles away by slow boat.

In retrospect, it was an interesting trip. For instance, I found that many of the hands, roughnecks and roustabouts, are dry-land types which maintain homes in Mississippi, Alabama, and Texas, well away from the coast.

While they complain about being away from home a week at a stretch—even more on some rigs—the week-off schedule has great appeal, as does the hefty paycheck. And they appreciate the relief from big room-and-board bills which are the bane of their land-based contemporaries.

Our first try to reach the rig started shortly after dawn when a helicopter lifted us off the pad and turned south by west, dodging between the rain showers. But dodging takes time, and time means fuel; so an hour later we turned back to the barn in hopes of clearing weather in the afternoon.

Clear it did, the brilliant sun chasing the clouds and turning the chopper into a fluttering sauna. Aloft, however, an ominous darkness was apparent in the area of the rig. As we drew closer, the darkness grew deeper and the showers started again, hanging about us like inspectors tip-toeing across the ocean.

Water spouts were the last straw. Although the pilot made a cautious effort to edge around them, radio reports said there was little chance to reach the rig, and there was no disappointment when again the compass pointed north.

By 6 p.m. we were back on the ground, our ears deafened and fannies sore from 5 hr of screaming jounce and jiggle. Mr. Sikorski's invention is wonderful, but the oil-field version incorporates neither quiet nor comfort.

Neither does it move at night. But crewboats do.

I love fishing, the surf, the sea, and boats of all kinds. I also suffer from mal de mer and have been known to throw up at the mere loosening of a spring line. So it was with mixed emotions that I boarded the boat that night, faced with an 8-hr run through quartering 5-ft swells.

It was midnight when we cleared the sea locks and it was 6 a.m. when we reached the rig. I do not choose to recall those 6 hr. Nor would I repeat them for a Pulitzer Prize with a Croix de Guerre thrown in.

Journalism is a great career, but sometimes it asks a lot.

Circumstantial evidence

John L. Kennedy | **January 10, 1972**

"You sure picked a bad time to go to the North Sea." That's what everyone said when I left Houston in mid-October to spend a month talking technical turkey with North Sea operators and contractors.

But I didn't look at it that way. One thing I really wanted out of the trip, besides a bundle of dope for the special section in this issue, was a real, honest-to-goodness "experience." A face-to-face encounter with the awesome sea. Something I could spring when hotel-lobby conversation turns to close calls. Something impressive I could use in this column.

All the way to London, I daydreamed of clinging to a storm-battered crewboat, then being plucked from a Block 22/6 grave by a courageous helicopter pilot. Through it all, I'd be snapping award-winning, Journal-cover pictures, certain to win me a handsome raise.

Since I already had a couple of offshore visits lined up, it looked like a sure thing. If I couldn't have my dream disaster in the North Sea, where could I have it?

Unfortunately, I didn't get the opportunity to abandon any crewboats. I didn't even see any giant swells.

I didn't get a smell of danger in 4 weeks!

The only shred of first-hand evidence that the North Sea is difficult I picked up on a visit to an offshore platform. There I got a hint the sea could possibly be trouble. Though the day was beautiful, and the sea looked like a green slate counter top—not a whitecap in a carload—it took a skilled operator some time to snatch two small doghouses and a food container from a heaving supply boat onto the platform.

This is virtually the only rough weather data I confirmed.

However, I've been studying all the bits of information that support the contention that the North Sea is rough. And some of it is pretty convincing.

For example, would a smart operator pay to have 30 days' storage space for mud and cement built into an offshore platform if it wasn't needed?

Would contractors build mobile drilling rigs to withstand 75-ft waves if these conditions weren't possible?

Would rigs report 30-ft seas to the head office just to impress the Journal's drilling editor?

Still . . . I haven't seen a "bad" sea. Never had the thrill of thinking the end is near.

The evidence is overwhelming that the North Sea is the roughest environment the industry has faced.

But in my case, at least so far, it's all circumstantial.

The picture profession

Carl Hoot **January 24, 1972**

As Journal staff photographer, my assignments provide many opportunities for travel. To far-off places like the West Texas desert. To the boondocks of the Gulf Coast. With luck, an occasional safari to Kansas.

To be sure, the Journal does publish many fine pieces from really far-off places. But most of the time, these exotic trips are made, stories written, and pictures taken by the editors. Photographers hate editors who do that.

So when somebody mentioned sending the real OGJ photographer to Cherry Point to illustrate our report on ARCO's new superduper refinery there, I sat up and paid attention, even though I had no idea where Cherry Point was. It turned out to be in the State of Washington near the Canadian border. It was described as a pleasant enough spot if you didn't mind rain falling about 24 hr each day.

That called for special planning. Photographers always have to plan to stay ahead of unfeeling editors, who send them north in winter and south in summer.

To start things off right, the trip would be made in June. Side trips would certainly have to be considered. Side trips are necessary to prolong the main trip and to show that the photographer's role in producing a comprehensive pictorial coverage is complex and pretty darn important. Plans were made to visit nearby Mt. Baker and all the Puget Sound islands. With luck, Banff might be included. A day or two might even be allotted for pictures at the refinery itself.

The trip didn't really work out that way, of course, since editors hate photographers. Sunny June's scheduling dragged out to Shady November's reality. November, you understand, is "between seasons"—that is, between summer rains and winter rains. During this period, it turns out, it keeps on raining.

One of my best pictures of the refinery is either sunrise or sunset, depending on how you look at it. It was taken 15 min after the sun came out for the day, and 15 min before it went down, also for the day. It stopped raining just long enough for me to take the picture.

Trouble is, those confounded editors are now using the picture to claim I was frolicking around up there in Florida-type weather all the time I was freezing my eye-lashes.

You can see why your average Journal photographer doesn't care for your average Journal editor.

Notes from a Halbouty watcher

Howard M. Wilson **May 8, 1972**

In 15 years of covering Mike Halbouty speeches we've become an expert.

We used to catch Mike on our Texas beat. Now we catch him on his periodic forays to the West Coast.

If you've never seen or heard one of the oil industry's most persuasive platform artists, you have denied yourself one of life's most gratifying—but unnerving—experiences. Also you're not getting out to enough meetings.

The moment Mike moves to the microphone, the banquet hall is his. Ramrod straight, exuding confidence, eyes glaring across the audience, he speaks just like he looks. No shilly-shally. He drives pell-mell to the center of the coconut.

It's time, he says, for the oil industry to get off its backside, quit taking insults, and attack!

"And I would like to know why we don't start fighting back now! Not tomorrow—not next month—not next year—but NOW!"

You feel it's time to quit the banquet hall and march on City Hall.

But Mike can vary his pace. Suddenly his voice softens, almost to a whisper. He warns that this government, in bringing the domestic oil industry to its knees, may be writing its own epitaph:

"Here lies a country, at one time the greatest in the world, which was relegated to a fourth-rate power when it became dependent on foreign sources for its greatest energy need—petroleum. The foreign powers which controlled the flow of petroleum into this sick land controlled the air above and the seas around it—until the United States expired. The short-sightedness of a few demagogic politicians during the decade of 1965-75 A.D. was responsible for the tragedy."

His tone still soft, his audience dazed at the prospect of oil's demise 3 years hence, Mike calls for prayer—a prayer that those who militate against our vital industry will comprehend the need to change their ways:

"Let us pray that this comprehension comes to them now—and not tomorrow—for tomorrow may be too late."

The speech over, the audience rises without hesitation to give Mike the standing applause he has earned.

To the reporter, it's a frustrating assignment reducing to cold type the words spoken on specific issues by the oilman-geologist from Houston.

The fact is, of course, that what Mike Halbouty is saying is far more important than the way he says it.

With a fervor that reflects not only his own restless nature but also the agonizing struggle which oilmen face in today's antagonistic world, Mike is goading his fellow professionals and businessmen to move out of their foxholes—"the trenches of stupefaction"—and shout aloud their side of the story.

If he makes his pitch with the flamboyance of a Jennings Bryan, we've no quarrel with that. He does have a vital message for the industry. And, even in cold type, we'll help him deliver it next time he comes to California.

A "misteak" brews a storm

Frank J. Gardner **July 24, 1972**

A couple of months ago, we filled this space with an item headed "The life of a columnist."

In it, we mentioned the matter of letters from the readers, and said "Just let us make a misteak, and here come the letters!"

Well, that "misteak" really brewed a storm. We were deluged with letters calling our attention to it.

The "misteak," of course, was intentional—our feeble attempt at humor. But reader reaction was amazing. Some of the letters recognized the situation for what it was, while others indicated we were unaware we'd made a real mistake.

May we share these with you?

First one to arrive was addressed not to us, but to our cohort Larry Auldridge. "I refuse," it said, "to be suckered into increasing your buddy's fan mail by such a scurvy trick!"

"I just wanted you to know," read another letter, "that I'm not going to join the rush of people pointing out your misteak in the May 29 column."

"Now, honestly," said another, "did you do that purposely to get a letter or two, Mr. G? Continued success to a man whose work is not a drag. How very, very refreshing!"

Another reader simply clipped the column, circled "misteak" and penned in "Spelling or pun?"

On a postcard: "I give up, Frank, was the misteak deliberate?"

From the executive vice-president of one of our leading independent producers associations: "Question: Is that a typo, or is it bait for letters?"

But our favorite of all the letters was a handwritten one, rather long, which had the notation "8:03 a.m." after the return address.

"You are so right," it led off, "in knowing what will make people put a pen to paper. All of us are so proud to find a misteak in the other fellow's work that we sometimes put one in on purpose, just to make him feel good.

"And, too," the letter continued, "just to write can make a fellow feel good. But also, just to know that what one has written has also been read can make one feel even better. And so, I feel very good just now for having broken the ice—and my pencil—in penning these words—my first-ever letter to an editor.

"Too, I hope that this little effort will make you feel good, too, especially since I haven't underlined the misteak in your column.

"And now," the reader concluded, "having gotten this off my chest, I think that the day to follow will be a little better.

Thanks for giving us a lift."

With friends like that, who needs misteaks?

European expose

Max L. Batchelder **July 31, 1972**

Journal editors are the best-traveled people you ever met.

To try to catch up, my wife and I recently took off on a whirlwind 10-

country tour of Europe. Our guide was an Italian citizen of Austrian descent attending a university in Switzerland and conducting tours in English.

At least he thought it was English. Along the Rhine, for instance, he told us about the Thievery Knights, who turned out to be Robber Barons, and quoted an Austrian emperor as declaring flatly, "In my empire, the sun never sets up down."

I never did find what he meant by saying of Marie Theresa: "She was able to defeat herself at all times when attacked."

I learned to let sleeping dogs lie. When he explained that Venice was built on thousands of wooden pilots, I never questioned it. When he pointed out that women teachers in the Middle Ages had to have screams in front of them to keep trouble away, I let it go. When he said that Rome has a circle highway around so you don't have to go through town unless you have to go through town, I said nothing.

I admit I was nearly stirred to action when he described grilling a chicken Italian style: "They put carrots in the back-side and parsley on top." I said nothing, but never ate another carrot the whole trip.

A guide I can cope with. Other things get me down. I needed to go to the restroom in Italy, so I marched into the proper door and found there only a girl seated before a cash register.

She raised her eyebrows and asked, "Pedicure or toilet?" I hadn't given it all that much thought, but I sure didn't want a pedicure. She nodded, collected 10 lire, and handed me a sales slip, motioning me toward a door. I opened the door and stepped in. I was in a small room with six smooth sides, and nothing else.

I was nonplussed. Then, with a jerk, the room began to rise, and I realized I was in an elevator. The elevator stopped, a woman pulled the door open, gestured me out, glanced at my ticket, and ushered me down a row of stalls to an empty one. I have never in my life seen such pomp and ceremony connected with going to the bathroom. And, God willing, I never hope to again.

But even this pales beside what happened in Paris, 3 days before the trip was to end. My wife and I went into a restaurant and were given a menu. My eye caught something.

"My God, look at this, Chris! " I whispered, pointing shakily at one listing. "Poison soup. Says right on there, plain as day. Poison soup."

My wife eyed me patiently. "No, Max, That's poisson. Soupe de Poisson. Fish soup."

"You're kidding."

Well, she had to prove it, so she ordered me some and it came and I ate it, and I got sicker than a dog and never got out of bed for the entire 3 days we were in Paris.

I guess I know poison soup when I see it on the menu.

From bear to bull

Frank J. Gardner **August 7, 1972**

As you can see from the lead story in this issue of the Journal, the North Sea is fast shaping up into a major oil province, and this reporter cannot overemphasize its future importance.

'Twas not always thus. The first time we ever heard anyone mention the possibility of oil or gas in the North Sea was in February 1963. Our editor, George Weber, invited us to go with him to a banquet in Tulsa—a black-tie affair. The speaker was a geologist, Chris Dohm, then with Pan American Petroleum.

Dohm's theory was that the Permian sands that were gas-productive at northern Holland's big Groningen field should logically sweep across the North Sea to Britain.

Leaving the banquet, we scoffed to the editor, and using a cliché, offered to drink all the oil or gas they ever found in the North Sea.

For the next 3 years, as dry hole followed dry hole in the sea, we followed that bearish approach. When British Petroleum found gas at West Sole at the end of 1965, we said it was a freak occurrence.

But as those giant gas fields grew off Britain's shores, we began to change our thinking. Maybe there was something after all.

At the time, we had an astute correspondent in Norway, who suddenly began sending us stories of a highly secret oil strike in Norwegian waters. At first, we ignored them—no rumors for OGJ. But as his stories became more factual and insistent, the rumor became too strong to resist. So we handled the item in a "Watching the World" column in a rather cagey manner.

What happened? Our first rejection slip from OGJ! "Too optimistic."

But 2 weeks later, in the face of ever-stronger rumors right from the derrick floor, Tulsa decided to print the column after all.

Then, one of Phillips' partners in the Ekofisk drilling could hold itself in no longer. Belgium's Petrofina dropped the word that a 7-billion-bbl oil field had been found.

That's all we needed. We wrote a major lead story for OGJ, and pulled no punches on the great significance of the find. But the headline that Tulsa slapped on it made even us cringe—"Huge North Sea find has entire oil world vibrating." We really hadn't been that bullish about it!

But it was true. Two issues later (OGJ, June 8, 1970, p. 81), we carried a story quoting Phillips' Pres. John Houchin as saying "It appears that we have discovered a giant oil field."

We relaxed. And as Tenneco announced at about the same time that it had found a big oil field in the North Sea off Holland, we knew that a new era had arrived.

If you don't believe it, just read on!

The intellectual beachcomber

L. R. Aalund **August 21, 1972**

Houston and the sizable chunk of world refining and petrochemical capacity near it face a very large problem.

It seems that Trinity and Galveston bays, which serve as this area's life line to the sea, are in the paths of large deltas advancing seaward. The reason is that only within the past 1,000 years has the Trinity River adjusted to its base level. Now this underachiever and the nearby San Jacinto River, given sufficient time and a constant base level, should fill the bays with sediments.

But don't trade in the boat on a horse yet. The U.S. Corps of Engineers may have a thing to say about this and it is a geological process in which time does have a way of stretching out.

And above all this is the interpretation of the OGJ Refining Editor—who should maybe stick to refining—of a fact gleaned from a fascinating guidebook, "Recent Sediments of Southeast Texas." This was published by the Bureau of Economic Geology at the University of Texas at Austin.

But this guidebook, which was written and compiled by geologists in Shell Development Co.'s exploration and production division, is one of those rare works that has something both for the scientist and for the layman.

Though it could be used as a guidebook for freshman geologists, it is fleshed out with meaty papers, including a detailed résumé of the quaternary geology of the northwestern Gulf of Mexico.

Of course, Shell Development didn't study meander belts, barrier islands, etc., to present an interesting booklet. R.J. Le Blanc, one of the contributors, says that a knowledge of these processes can give clues to new oil.

But the book does open the eyes of the average beachcomber to things he has never really seen. And having Le Blanc along for a walk down the beach would be most interesting.

For example, near High Island, Tex., he has the first oak tree spotted that is a forerunner of the vast Chenier plain, which is the main feature of the Louisiana coast.

Cheniers (local French usage meaning oak grove) are old beaches that are stranded inland as the marshes move seaward. The old beaches are high enough to give secure footing for the ancient, beautiful oaks in that part of Louisiana. Had the prevailing tree there been pine, Le Blanc says with a twinkle, we could call them "Pineiers."

Chenier crops up in many place names in South Louisiana such as Grand Chenier and Chenier au Tigre. However, one crossroads settlement south of Creole, La., bears the name Oak Grove.

"Originally settled by Yankees, no doubt, who thought they hit on a unique name," muses Le Blanc.

Voyage to the bottom of the sea

W. B. Bleakley **October 2, 1972**

We Journal editors will go anywhere and do almost anything to get a story.

We have a certificate that shows just how far we'll go.

We told you about the Lockheed Petroleum Services Ltd. underwater production system now operating on a Shell Oil Co. well in August. But at that time, the diving capsule was under test, not making routine dives.

And 3 weeks from this issue date, we'll describe how the diving capsule was used to install the wellhead in a shirt-sleeve environment 375 ft under the Gulf of Mexico—a new water-depth record. But let me tell you about my voyage into the ocean depths to gain first-hand knowledge of this operation.

We climbed into the capsule through the bottom hatch, found seats, and prepared for launch. Just as the hatch was dogged in place, I felt pressure in my ears.

The capsule commander, Lockheed's Larry Gentry, said, "We're going to increase the capsule air pressure . . ."

"Could you speak a little louder, please. I can't hear for all this pressure in my ears," I said.

He continued. "Reason for boosting the pressure is so we can open the hatch to evacuate while floating, if necessary."

I gulped and checked the fasteners on my life vest.

The launch brought with it a lot of swinging, bucking, and heaving of the capsule as it was lifted clear of the deck and swung sternward over the water. Once in the water, the capsule bobbed on the waves like any other floating object.

Gentry started the hydraulic winch and we began our descent. The small glass ports showed we were indeed underwater as curious fish took turns staring back at us.

Down to 25 ft we could still feel wave motion, but it became less and less noticeable.

"Passing 50 ft," Gentry said into his microphone to the control room aboard the Coastal Cruiser. His TV monitor could not yet pick up the wellhead cellar still 325 ft below us.

Suddenly all was quiet, almost, and all motion, seemed to stop. We were below wave action, descending slowly.

"What's that noise?" I said.

"What noise?" Gentry asked.

"Sort of a boney, banging noise, like somebody's knees knocking together," I said.

"Oh, that. It is, and they're yours," Gentry enlightened me, and I clamped them together for the rest of the trip.

By air, land, and sea, we Journal editors go everywhere. But this is our first venture under the sea. The way the industry is moving, we'll guess it won't be the last.

Yo ho ho and a barrel of resid

Leo R. Aalund **November 6, 1972**

"Ah, this is a sweet spot, this island.... Why it makes me young again. I was going to forget my timber leg, I was."

When the mutinous scoundrel Long John Silver spoke those words to Jim Hawkins on deck the Hispaniola off Treasure Island, he had the glint of buccaneer's gold in his eyes.

Now two centuries later, honest seafaring men are again looking at maps of the West Indies and soundings of the waters. They, too, like what they see—waters deep enough for 300,000-dwt tankers and prime refinery sites.

Sadly, perhaps, the fabled Spanish Main now evokes more than thoughts of gold, rum, and the skull and cross bones. It means a growing refining establishment.

For different reasons, this development is disturbing for many in the U.S. and in the islands.

"Stinking clouds over the Caribbean," was one headline for a recent, widely circulated Associated Press feature. Replete with pictures, it decries the environmental impact of refineries and petrochemical plants on the Virgin Islands and Puerto Rico.

But natural beauty and clean air haven't made life easy for the many poor inhabitants of this necklace of emeralds mounted on pearl that stretches some 1,500 miles across a blue sea. The question is, will tourism or industry feed the most people?

In the U.S., some in government and industry are worried about the exodus of U.S. refining capacity to those shores. National security and balance of payments are the central issues. Memories of the wolf pack in the Caribbean are still vivid.

Then, too, there is Communist Cuba.

Politically, most of the Caribbean is in its infancy. Practically every governmental relationship known is being tested.

They range from new independent island nations, like Trinidad and Tobago, to pure and simple possessions, like the Turks and Caicos Islands. In between, there are the Bahamas with internal autonomy and headed for full independence.

Trade and the affairs of the West Indies have long concerned the U.S. East Coast. There are even striking parallels in history with the current importation of resid.

In the early 1700's, the New England colonies and New York were importing rum and molasses (to make more rum) from British and foreign islands in the West Indies. The British planters (domestic producers) got Parliament to slap a prohibitive duty on foreign rum and sugar so their countrymen would have to buy their stuff. Then history deviates slightly. There was no Department of Interior then, so the colonists started smuggling the rum in.

Gold on the sidewalks of New York

Frank J. Gardner **December 25, 1972**

This incredible story points up two realities—the great number of oilmen in New York City, and the widespread readership of The Oil and Gas Journal there. It also destroys a myth—the coldheartedness of New Yorkers.

For a bit of background: Each Christmas, OGJ presents its employees with gold service pins, marking 5, 10, 15, or whatever one's years of service in multiples of 5.

For the men, these are presented in the form of small square tie-tacks, about 1/2-in. across. They carry the initials OGJ and (in our case) FJG, with the number of years of service in the center.

Now for the story.

A couple of months ago, we arrived at the office on a cold Monday morning. Coats off, shirtsleeves at the ready, we said "good morning" to our alter-ego Royal typewriter and immediately noticed our necktie was blowing freely in the breeze . . . no tie-tack! Oh well, must've forgotten the darn thing this morning.

Not 5 minutes later, our Mrs. Ruth Schmalz hands us a pink slip recording a phone call that had arrived before we did.

"Mr. Pokrassa called. Please call him at 755-5530."

Mr. Pokrassa? Who is he? Don't know him.

But we returned the call.

"Mr. Gardner? Yes, you don't know me, but I think I've found something that belongs to you. I was walking down 57th St. this morning, and my eye was caught by a bit of gold glinting in the sun. It was in the middle of 57th, but I picked it up and looked at it. It has the initials OGJ and FJG engraved on it, and a figure 15."

"Don't tell me you found my missing tie tack!"

"Yes, I think so. You see, as a long-time reader of your magazine, I'm quite familiar with the OGJ emblem, and I knew the FJG had to be that guy who writes 'Watching the World'!"

"The pin part," he added, "has been broken off. A car probably ran over it. So I'll just tape it to a card and drop it in the mail to you."

Thanking the man profusely, we hung up the phone in disbelief. A 1/2-in. square of gold, lost and found within half an hour on the sidewalks of

New York? Incredible!

But that 's not the end. Another half hour, and Mrs. Schmalz tells us Mr. Pokrassa is waiting to see us.

"I thought I'd better bring the pin right over. It might get lost in the mail," the man said.

He handed us an envelope. In it were the tie-tack and a short typewritten note. "Even we in the advertising field know the importance of something with sentimental value," it read.

Thanking the man again, we shook hands. But as a parting shot, we asked "Where are you from, sir?" (Surely from the South or West, we thought.)

"Right here. Native New Yorker," came the reply.

Once an oilman

Carl Lawrence **January 1, 1973**

Roy Ellison, circulation manager for the Oil & Gas Journal, says oilmen have a strange disease. It gets in their blood and just won't go away.

"I've never seen an industry like the oil business," Roy says. "Normally, a man retires and forgets about the industry he's been in. But not oilmen. When they retire, they want to stay abreast of the industry and want to keep right on receiving the Journal every week."

This presents problems for Roy. He faces more regulations than a wildcatter trying to drill in the middle of a wildlife refuge. The Audit Bureau of Circulations (ABC) spells out just who can and who can't buy an OGJ subscription at the regular industry rate. And our advertising department sells accounts based on Roy's ability to come up with a good ABC statement.

These retired oilmen who want to stay informed present Roy with another problem. Just like any other businessman, Roy faces rising costs. He figures it costs $30/year to produce and mail a Journal to a subscriber. When you're only charging a domestic rate of $12/year as Roy does a regular OGJ subscription, you're obviously not making money—you're losing $18/year/subscription. Of course, if Roy does his job well, then our ABC statement convinces advertisers the Journal is the place for them, so we make up in advertising revenue what we lose in circulation costs.

But what about the retired oilman? Where does he fit in Roy's ABC statement. Well, he fits in an expensive spot. He's carried as a total loss when it comes to selling advertising. But Roy's a nice guy, and he'll keep our retired readers on the subscription list for a price—$32.50/year. Roy says this is his breakeven price. He doesn't try to make money on the retired oilmen, but he does try to avoid losing any money. When he loses $18 on every regular sale to an active oilman, he can't afford to lose money on the retired oilman.

Explaining this is where the rub comes in, especially if the retired oilman happens to be one of those subscribers who had been reading a company-paid sub for years. And, of course, Roy has learned oilmen—retired or otherwise—aren't bashful about expressing themselves, especially when they think they're getting the short end of the stick.

Our old friend, Sam Bowlby, who retired a few years ago after an illustrious career with Shell Oil Co., had some choice comments to make when he discovered his OGJ sub was no longer in effect. Sam said that we were a sorry bunch, that he had been a faithful reader for 40 years, and that he had hardly left the Shell Building in Los Angeles before we canceled his sub. Well, it didn't happen quite that fast, but Sam's sub was dropped. However, Roy picked it up again, after some negotiating.

Sam's like hundreds of other retired oilmen who want to stay abreast of an industry they made a contribution to. He wants his OGJ weekly. The Marines used to have a saying: "Once a Marine, always a Marine." Roy Ellison says the same thing about oilmen. Once an oilman ...

Russian LNG specter achieves the impossible

Gene T. Kinney **January 15, 1973**

One Sunday morning recently, members of the Washington oil and gas lobby awoke in a panic.

They opened the morning paper and began to read. "At current low prices, producers have very little incentive to look for gas. . . . The present wellhead price of gas in this country could be tripled and, even adding transmission costs, would be far cheaper than the Soviet imports.... The consumer in this part of the country has a choice between the present low price and shortages, or a higher price and an adequate supply."

Who said that? Not IPAA, not AGA, not API, not Ingaa. It was the Washington Post, for years the country's most reliable baiter of the oil lobby.

The long-suffering lobbyists had scarcely recovered from the first assault when the Post did it again. It ran a second lead editorial on the same theme. Both were model presentations of the case for using market forces to balance supply and demand.

"There is a tendency in our part of the country to assume that whatever is good for the oil-and-gas lobby must be bad for the rest of us," the Post confessed. "Deregulation of natural gas is an exception to this rule. It will mean more money for the gas industry, but, much more important, it will mean adequate and reliable supplies for consumers."

This powerful expression of support from the federal Government's hometown paper stirred fear in the hearts of oil and gas reps. They didn't know whether to take credit for the Post's turnaround or to look for another job. The one thing lobbyists can't survive is being loved. If the Post's view should become contagious, they would be out on the street—along with two-thirds of the Nixon administration. They would be in the same boat as that bereft bureaucrat in the Bureau of Indian Affairs whose Indian died.

Lobbyists feel most secure when being abused. They understand the current campaign of a Delaware newspaper, which has a three-point program aimed at those who want to build refineries and tanker terminals. It is: "(1) Keep the bastards out. (2) Keep the bastards out. (3) Keep the bastards out." (To which the producing states respond with their old battle cry, "Let the bastards freeze," updated to add "in the dark.")

If the Post's attitude catches on, the lobbyists may have to leave Washington and move to Delaware. If the other paper prevails, we may all freeze to death in the dark.

Meanwhile, you can thank the Russians for doing the impossible. They have made the domestic industry look good in the nation's capital.

Nice going, Dick

John P. O'Donnell **January 22, 1973**

Back in 1948, a group of pipeline contractors met in the Mayo Hotel in Tulsa and formed the Pipe Line Contractors Association. This week the PLCA is celebrating its 25th anniversary at the Boca Raton Club in Florida.

The contrast in meeting sites reflects, in a way, the changes that have taken place in the contractors' estate in 25 years. The changes have made those years an eventful quarter of a century.

If someone had told the contractors at the 1948 meeting that they would do the huge volume of business they did in the mid-1960's, they would have questioned his sanity. And they would have done the same to anyone who had told them in the mid-1960's that they would do as little gas-line business as they are doing in the early 1970's.

When the contractors joined forces, the fantastic growth of natural-gas pipelines was just beginning. The 10 largest interstate gas pipelines operated 31,910 miles of pipelines in 1948. In 1971, the top 10 operated 134,757 miles. And all those miles, and many others, too, were put there by the contractors.

"Big-inch" construction in the late 1940's was "over 25-in." By 1948, interstate gas lines had laid 4,300 miles of it. Now "big-inch" is 30-in. and up. At the end of 1970, they had 40,950 miles of it in operation and almost 1,000 miles of that was 42-in.

Since they have been laying more miles of bigger pipe, it would appear that it has been all beer and skittles for the contractor. It hasn't been. Six of them laid the 1,045-mile, 36-in. section of Colonial in 1962-63. Only two are active today.

The mortality rate among the contractors has been high. They are in an acutely competitive business with an inordinately high risk factor. One of the risks they can temper is labor relations. This is where the PLCA comes in. Its first job is to represent the contractors in union negotiations.

Relations between contractors and labor have not always been in an aura of sweetness and light. They were strained most severely in the late 1960's when there were more miles of line to be welded than there were welders to weld them. That had to be the low spot of an otherwise relatively harmonious relationship.

Contractors have, pretty much, operated strike-free and, almost always, have been able to get the men they need when they need them. Much of the credit for this rather happy state of affairs must go to the PLCA.

Companies that make up the PLCA have changed greatly with the years and so have the members of its labor committees. Through all its 25 years, there has been a single continuous influence in the association. For that reason alone it must be regarded as a major factor in the success of the PLCA.

That influence is Richard A. Gump, managing director and the guiding element in the PLCA since its inception. You see Dick's name on the association letterhead with those of other officers and directors, and you hear a brief report from him at each annual meeting. Other than that, he's quite inconspicuous, apparently believing, as most of us were taught, that "actions speak louder than words."

The electric cookie

Leo R. Aalund **February 12, 1973**

Electric cookies smell delicious. This is the first thing you notice after walking from the plush reception area of the local "electric living center" through the auditorium and into a gleaming all-electric kitchen. And since it is Friday and about lunch time, there is fish baking, too.

In the auditorium, a group of ladies is circled up for some knitting and chit chat. In a little while, they will be sampling some of the dishes and perhaps get an expert demonstration of electric cooking.

This editor didn't get a cookie, though the friendly home economist who helps run the place would surely have given him one if asked.

But that's OK. Electric cookies are a symbol of one of the weirder aspects of the so-called U.S. energy crisis. They would stick in some people's throats.

Though the recipe is the same, it takes a lot more heat to bake them than the old-fashioned kind. Not that electric ovens are so terribly inefficient, but because "the light company," as they lately prefer to be called, has to burn about 10 BTU to get 3 into the kitchen. Engineers know this, but they aren't much in the way of cooks.

Of course, not all of the huge volume of natural gas that many utilities burn is used to generate electricity for the kitchen. There are some important things that only electricity can do well. This poses a special problem for "the light company" in places like Houston, the most air-conditioned city in the world. The summer generating peaks look like Everest, but the winter valleys mean slumbering generators.

So here and in other areas, somebody got the illuminating idea to push space heating, clothes drying, and egg frying with electricity. The wasted BTUs go into a creek, pond, or bay. Swimmers don't much care for this, but for the marine life it's supposed to be great. Why, with research, someday it may be possible to domesticize a colony of piranha fish from the warm waters of the Amazon.

Home and apartment builders, tempted by some nice carrots from utilities and appliance makers, jumped at the idea of electric heating. In Houston, for example, they have literally stampeded. Roughly 70% of all new apartments going in—and there are plenty—will use electric heating. So instead of heating at least two apartments with the original BTU, these folks will be heating one.

What energy crisis? To paraphrase Marie Antoinette, "if they don't have fossil fuels, let them use electricity."

OGJ makes Congressional Record

Gene T. Kinney **April 16, 1973**

You can learn a lot from reading the Congressional Record.

On Mar. 6, you could learn from Rep. Bertram L. Podell (D-N.Y.), one day late, that the day before was the start of "National Procrastination Week." In honor of the event, he put off his remarks on the subject until a later time.

Later last month, the Record also disclosed that Sen. William Proxmire (D-Wis.) spent $2,758 on his famous hair transplant, and counted it a medical deduction on his income tax return. You could also find out, a couple of days later, that as a result of his admirable self-disclosure, and the public outcry, he had had second thoughts about taking advantage of that tax loophole. He ponied up another $1,000 to the tax collector to set things straight.

You can also read a lot about the energy crisis in the daily chronicle of everything that's said on the floor of the House and Senate, and a lot that isn't. The more the energy crisis becomes front-page news, the more people in government turn to our magazine to find out what's happening—then use it to tell others.

As a result, we often see our work quoted, or reprinted at great length, in the Record, which is the eventual repository of all the country's wisdom. It is with a mixture of pride and humility that we find ourselves mentioned alongside the New York Times, the Washington Post, and the Christian Science Monitor. The day's Record may bring the latest expose or punditry of James Reston, Jack Anderson, Joseph C. Harsch, and Gene Kinney.

Gene who?

Yes, we find the Washington Editor and many others from the Journal staff are quoted frequently to prove one point or another. We are mentioned about as often as any other magazine, because energy is such a hot topic and congressmen turn to the "bible," as the Big Yellow Book is called in the Record and elsewhere.

We are pleased to have a role in informing government officials and the public. We're also delighted that they use our work to spread understanding. However, we sometimes wonder whether our innocent words aren't cited to prove a point other than the one we had in mind.

This seemed to be the case when Senator George McGovern (D-S.D.) pounced on our comments last month. We had suggested there might be room for improvement in oil-import policy, which didn't exactly spill any state secrets.

By and large, we like to think we are quoted not because somebody agrees or disagrees with our opinions. We hope it is because they consider our information reliable and useful.

Whatever the reason, we're glad to have a small part in keeping the Record straight.

Parlez-vous franglais?

Frank J. Gardner **May 7, 1973**

The French are bent on "purifying" their language by purging it of all those ugly American words that have infiltrated it since World War II.

And oil terms are no exception. In the Official Gazette published on Jan.

18, two separate lists of oil wordage appeared. One is obligatory for use in all government and public-administration publications, and the other is "suggested" for use by all media. There are 25 terms on the obligatory list and 50 on the suggested list.

From now on, Frenchmen are obliged by the government to avoid using the word "blowout." Correct term will be "eruption" (with the correct French pronunciation, of course—erupsyon).

And please—don't say "derrick" any more. It's to be "tour de forage" (drilling tower to us).

If you're tempted to use "workover," don't. The correct word from now on is "reconditionnement."

Refining terms really become exciting. For "cracking," it's to be "craquage," which is done, naturally, in a "craquer," not in a "cracking plant."

Even that old favorite "know-how" won't be proper any more in France. It is henceforth "savoir-faire."

Some words, mostly those of Latin origin, don't change except for the pronunciation—maintenance, completion, pipeline (peepleen), management.

On the second list we find some real jawbreakers—appareil de forage instead of rig, forage sauvage for wildcat (we like that one, and its relative, le puit sauvage, for wild well), colonne de production for tubing, and how about craquage catalytique for cat cracking.

Some of the French oil terms are positively sexy. Don't you twitch when you see "esquichage" instead of "squeeze job?" Or "en service" instead of "on stream?" Then there's "huile de schiste" for "shale oil."

You geologists, take care. Henceforth, it's "pergelisol," not "permafrost," "schiste argileux" for "shale," but "compaction" remains the same, provided you pronounce it "compak-syon."

The French will have a never-ending job in their flight from franglais, however. As one of our friends in Paris writes, "New words are born every day. What can they do with 'bridging crude' and 'phase-in crude'?"

Or Alyeska?

Reunion at the G.P.

William A. Bachman **June 18, 1973**

Those were the days. Remember when the Grand Pacific Hotel lobby at Bismarck was the unofficial oil headquarters for the now legendary Williston basin boom?

Just sit in the lobby long enough then and you could see every geologist, landman, or oil scout who pioneered that prolific North Dakota area in the early 1950's. Many of them have since retired or died or moved on.

Many of the survivors of the Williston play will return to the G.P. on June 24 and 25 for a "memory bash," a nostalgic reunion and the first held in the two decades since the Williston boom started. Meeting again at the G.P. has special significance, because this storied old hotel will be demolished soon.

A simple program has been arranged for the 2 days, starting Sunday, June 24, and featuring mainly informal sessions. The Sunday activity is capped by a 2-hr cocktail "hour" and beef barbecue, and the climax is Monday evening with another 2-hr cocktail "hour" and banquet.

Legends of the Williston started popping up immediately after announcement of the reunion, and scores more undoubtedly will be added by veterans coming to "bash."

One concerns the late Thomas Leach of Bismarck, former president of North American Royalties, named the "granddaddy" of the Williston oil play. He first leased the Nesson anticline for The California Co. with his partner, the late A. L. Fruh. A dry hole was drilled on the acreage before World War II within 2 miles of what was later to be Beaver Lodge field, the first in the state. He again leased the Nesson for Amerada in the late 1940's, and the discovery was drilled in 1951.

Then there was the division manager of a major oil company who pulled up in a cab outside the G.P. and asked the manager if he could put the cab fare on his hotel bill. The manager agreed but was a little taken aback when the cabbie said the fare was $181.50. The cab had come from Williston, 200 miles north. The fare, however, went on the bill.

Another time, two lease brokers in league with the G.P. desk clerk registered a slightly wounded wild goose into a room under the name A. Goose and then set up a series of buying appointments for Mr. Goose with other brokers.

At the conclusion of a Hawaiian luau, one guest ceremoniously took his exit through a door at the hotel which formerly entered on a balcony which had been removed. He fell to the sidewalk but picked himself up and walked to his room.

There also were some strange happenings outside the G.P. in those days. One broker in 1952 bought a steak at a night club and paid with a $3,400 check from that day's mineral deal. The manager gave him $3,393 in cash from the club's safe for change. At another night club, a geologist tipped the band $200 to play past the Sunday morning closing time. And the North Dakota Geological Society threw a free wild-game dinner for oilmen—1,000 turned out and ran up a $2,900 bill which took the society 2 years to pay off.

And there are probably many more ghosts of those exciting years which will be revived in the G.P. lobby next week.

A one-well story

Carlos Byars **July 2, 1973**

Little Caesar Oil Co. will never make Fortune's Top 500—or even the top 5,000. But I wish I owned it.

Granted, it's not much of an oil company. No refinery, no import quota, no pipelines, no service stations, and no headaches. It started small and has stayed small; just one good well on a 1-acre lease in the East Texas field.

Nolan Rushing of the El Dorado Rushings owns it, and, now and again, he drives down to Kilgore to take a look at Little Caesar and his other holdings. It's a familiar drive. He put the deal together over 40 years ago, back when the lights burned day and night in the supply stores and the All American restaurant was the place to eat in Kilgore.

Like a lot of other people, Nolan Rushing went to Kilgore in 1931 to get in on the boom. So did his cousin, Dr. J. S. Rushing, an El Dorado dentist who also owned a drilling rig. Another notable was a Princeton grad named F. W. Graves, who is remembered for having hit town in a chauffeur-driven limousine.

Nolan was single at the time and soon was going with a girl from Henderson. The girl's father, a local political figure, was buying and selling leases, one of which covered a 70-acre farm and 1-acre orchard.

The 70 acres soon were sold for cash and oil, but Nolan had his eye on the orchard and was able to make a deal—$3,000 for the acre to be paid out of one-sixth of any production.

To get the well drilled, Nolan arranged for the use of his cousin's rig and, for a half interest, Graves put up the money.

Nolan is supposed to have said later that all Little Caesar cost him was a $3 filing fee. That isn't quite correct, since he had to pay Hamp Snead, a Longview attorney, to incorporate the outfit. Still it was a bargain. Based on an hour's test, the well was assigned a potential of 735 b/d and carried a 20 b/d allowable for nearly 40 years. It's still producing, though on the pump and making water.

The name? Some oldtimers say it came from the name of a movie popular at the time starring Edward G. Robinson. Indirectly, it may have, but Nolan says he never saw the show.

Hamp asked me what name I wanted for the company, and I didn't know, so he suggested Little Caesar because all I had was 'a little lease and a little bit of money.'"

ARCO's bus plan

Howard M. Wilson **July 9, 1973**

As a veteran bus rider who has been trying to stave off the energy crisis and eliminate smog for the past 15 years, I am delighted to report that Atlantic Richfield Co. is now lending support to my single-handed effort.

ARCO, as you know, recently moved its corporate headquarters from New York to the new ARCO Towers in downtown Los Angeles and, consequently, is pulling 1,000 persons daily into the central city. These people drive to work from the far reaches of the Los Angeles Basin, thus contributing to smog and inordinate use of energy.

Living up to its reputation for being a good-citizen company, ARCO has developed a busing plan that may be unique in the nation. If it succeeds, others may try it.

Here's how it works. On the morning of July 2, seven buses started picking up more than 200 ARCO employees living 20 to 30 miles from the office and depositing them at the ARCO Towers at 7:45 a.m.—just in time to grab a cup of coffee before reporting to work at 8 a.m. Then at 4:40 p.m., 10 minutes after closing time, the workers boarded the same buses for home.

To add incentive, ARCO will pay $15 of the $40 monthly tab. The remaining $25 is a payroll deduction. Thus ARCO is spending some $3,000/month to decrease use of its own principal product, gasoline.

The program is a joint venture with the local bus company, which furnishes the air-conditioned buses and the drivers. Since the bus company requires 35 passengers per bus, ARCO is busy rounding up more non-ARCO riders to help meet the guarantee. This minimum rises in 6 months to 40 passengers—which means ARCO is betting on the come. The idea must catch on with others who work in the ARCO Towers area.

The bus routes stretch in all directions from downtown Los Angeles but principally northwest, northeast, east, and south. Lines going straight north and straight west failed to attract enough rides to warrant start-up of service at this time.

The buses make three or four stops at locations where riders can park their cars if they choose to drive to the bus line. If a rider misses his bus, he can use his ticket on a regular bus line.

ARCO hasn't figured out yet just how much gasoline business it will lose by this scheme. But we figure that 200 employees who drive 40 miles a day (and nearly all of them do at present), or a total of 8,000 miles, would be using 800 gal of gas a day. Subtract from that figure the fuel that seven buses will consume, and you have an idea of what one company is doing to save energy and reduce smog.

There's just one trouble with this scheme. One of those buses that failed to get enough riders would go right past my own bus stop on Verdugo Road in Glendale. That being so, I've had to cancel my request to the Oil & Gas

Journal, a good-citizen company, to subsidize my transportation at the rate of $15/month.

Buzzphrase generator—Phase II

W. A. (Bill) Bachman **August 6, 1973**

0. Interdependent	0. Social	0. Interrelations
1. Stringent	1. Protective	1. Attitudes
2. Viable	2. Mitigating	2. Concerns
3. Alternate	3. Ecological	3. Factors
4. Natural	4. Sensitizing	4. Values
5. Significant	5. Environmental	5. Parameters
6. Exceptional	6. Deteriorating	6. Effects
7. Contributory	7. Technocratic	7. Limitations
8. Stimulatory	8. Predisposing	8. Priorities
9. Enhanced	9. Facilitating	9. Surveys

Many Journal readers are aware of this magazine's deep interest in bridging any language gap that may arise with the introduction of a new discipline or technology to American life.

Back in 1967, Journal Editorial Writer Henry D. Ralph introduced in this column the now-famous buzzphrase generator enabling anyone to talk glibly about computer intricacies at cocktail parties or other gatherings (OGJ, Dec. 11, 1967, p. 61). It enabled many an oilman to live through the early computer era safely.

Now, the environment and ecology have grabbed everyone's attention. A new language has appeared surrounding the dogma and mysteries of the new religion. The uninitiated often are completely out of it at cocktail parties and buzz sessions.

But Don E. Brown, a Tulsa professional engineer, has devised another buzzphrase generator to fit this situation and is willing to share it with Journal readers.

Here's how it works, Pick any three-digit number, for example 678. Then refer to the phrase generator above. Select the word opposite 6 in Column 1, the word opposite 7 in Column 2, and the word opposite 8 in Column 3. Thus, you have "exceptional technocratic priorities," a good environmental phrase. Repeat process with any other three-digit combination for other phrases.

Brown wisely advises, "If you do not understand its meaning, don't worry, just fake it. With the use of buzzphrases as prolific as they are today, it will take an absolute master of the English lexicon to detect any lack of meaning."

"After you and your colleagues have familiarized yourselves with the phrase generator," Brown says, "it may be possible simply to state the selected three-digit number, automatically recognizing the three words associated therewith. By using the numerals as abbreviations much time and effort could be saved with no loss of communication."

Now, you're on your own. Happy environmental phrase-making!

Dining out

Leo R. Aalund **August 13, 1973**

During a recent assignment in London and the Hague, this editor for several reasons—jet lag, tight scheduling, etc.—wasn't always ready or able to eat when everyone else was.

Trying, for example, to find a cup of coffee and a couple of eggs at about 11 a.m. on a Saturday morning in a certain area of Kensington can be futile.

The hotel porter had reacted as if he had been asked for the loan of a pair of skis when queried if breakfast was still being served.

Well then, lunch must be being served. Another look of disbelief.

So, out into the sunny streets to try elsewhere. An elderly British lady with a shopping bag looked knowledgeable about the neighborhood.

"Excuse me ma'am, can you tell me where I can get coffee? "

"Why yes, there's a grocer's just across the street."

"No, I mean a cup of coffee to drink."

"Well," she said, "buy it and brew it yourself."

"You see, I am in this hotel down the street and...

"Oh dear, oh dear," she said as her hand went to her throat in a movement of shock. I could feel an invitation to coffee in a Kensington flat coming on from this kind soul and hurriedly departed with thanks.

But luckily these two cities have institutions that keep transients from starving—even on holidays.

In London, it's the East Indian Grocer. There's one in Earl's Court that probably never closes. The egg-salad sandwiches to go look the most familiar. But the madhouse of Earl's Court at night is not conducive to a relaxed meal. However, a slow walk toward the BEA terminal is pleasant, and you can hold your half-eaten sandwich unobtrusively by your side when you pass someone under a street lamp. It is slightly embarrassing, however, if you have a mouthful and are stopped and asked where Lexham Gardens is.

The reply, "I am from Texas," is the quickest way to explain the ignorance—and perhaps the egg salad—for two London ladies.

In the Hague, the answer to hunger is the Herring King. He has a little mobile restaurant. He is inside, and you are on the sidewalk.

The new herring from the North Sea are delicious. Have a couple right from the counter (pick them up by the tail and eat sword-swallower style). Then try a shrimp sandwich. A nearby bench by the darkened canal is a peaceful place to enjoy it. After that, go back for a portion of smoked eel or mackerel.

Dining out in Europe can be delightful.

A little something to keep out the cold

Ed McGhee **September 10, 1973**

We'll never forget the first time we visited an offshore rig owned by a French company. Wine was served at lunch.

Not very remarkable for the French, you say. And we admit that the sight of a roughneck washing down his food with something alcoholic would startle but few. What gave us pause was where he was doing it—not 200 ft from the rotary table.

Most drilling men have it pounded into them early: Alcohol and rigs do not mix. Get caught with booze on the rig, and you forfeit your job. Chances are too great that you'd lose a finger, an arm, or even your life. Or cost someone else his.

The experience on the French rig was scant preparation for the first time we visited a seismic camp in the Amazon jungle and found beer in the mess hall. Being assured that the practice is a time-honored one in remote areas didn't allay our suspicions about what mischief a seismic shooter might get into if drinking while handling his dynamite.

When we went to the Arctic this past summer to gather material for this issue of the Journal, perhaps we were already prepared mentally to see further departures from the ancient strictures. After all, our old grandfather had always assured us that a little tot of whisky would keep out the cold on a winter's day.

Our own previous experience in the Arctic left no doubt that the cold there does take an awful lot of keeping out. Besides, there's more to working on a rig in the Arctic than just the cold. There's the months-long blackness of the winter night. The drudgery of 12-hr tours day after day for 2 weeks running. The separation from family and friends. The cramped quarters. The unrelieved 24-hr throb of engines and pumps. It can be acutely depressing.

After experiencing this ourselves—even if only briefly—we didn't raise an eyebrow when an oil company's superintendent said, "We may have to bend the rules. Perhaps serve an open bottle of beer in the mess hall with the meal when a man comes off tower.

"When crewmen go that 2 weeks without a drink of any kind, some tend to overreact when they get back to Edmonton. Head for the nearest bar and booze it up bad. Their families don't like it. The men themselves aren't ready to get back on the airplane and go back to the rig after their week off.

"Maybe we can cushion this a bit. So they aren't so uptight when they start their days off. Anyhow, we will probably try and see how it works."

It reminded us of the story told by a veteran of a British Antarctic expedition. The cold was even more intense there on the bottom of the world, he said. Expedition food was spartan. The accommodations were meant only for survival. All a far cry from the tasty meals and solid quarters of the modern rig camp.

Yet no one on the expedition grumbled about the lack of liquor or anything else. How did he explain it? Simple. Every man was an unpaid volunteer.

We can't imagine manning an Arctic rig with volunteer unpaid roughnecks. So, we'll watch to see how the operators cope with the needs of the ones they've got.

Far North in summer

Howard M. Wilson **October 8, 1973**

If you've never undergone the rigors of a trip into the Far North, we suggest you sign on with Pacific Lighting Corp. of Los Angeles for a flying look at their potential gas properties.

We guarantee you 40-degree weather (plus 40, not minus) at Prudhoe Bay, 60 at Inuvik, and temperatures so warm at Yellow Knife that you sleep with your window open—that is, if you go in late summer or early September.

We had prepared for this 4-day journey into the Arctic by purchasing both an all-wool shirt and thermal underwear. We did find use for the shirt on occasion, but the long-handles stayed in the suitcase.

It was that kind of trip. Good weather, good company, spectacular scenery, and new insights into the search for gas in remote areas of the world.

Pacific Lighting flew us first to Kenai to inspect the site of its proposed LNG plant.

We spent the night in Anchorage and made the flight to Prudhoe the next morning. Our host thoughtfully provided an eye-popping view of Mt.

McKinley on the route north.

At Prudhoe, despite the overcast, we had decent weather for a good tour of the facilities. BP Alaska promised us a swim in their blue-tile pool if we come back next year. BP's new housing and recreational facilities are on the site at Prudhoe, but they won't be fully assembled until spring.

Both at Prudhoe and later at Inuvik, in the Mackenzie Delta, we heard Pacific Lighting's story on how the Arctic gas will be piped to southern Canada and the Lower 48.

This is a continuing story which the Journal carries as the project develops. But no story can tell it all. You have to see the wild terrain through which the pipeline must pass to appreciate fully what an awesome undertaking the gas industry is planning in its feverish drive to bring new supplies to market—and ease the energy crisis.

Our flying journey took us over one of the larger Arctic islands, Banks Island, then back down to Norman Wells and Sans Sault (by float plane) to see pipeline research projects, and finally to Yellow Knife, capital of the Northwest Territories.

In all, we traveled 7,500 miles.

We enjoyed one particular fringe benefit on the trip—collecting rocks for our rock-polishing tumbling machine.

If we find that rocks from the Cook Inlet, the Beaufort Sea, and the Mackenzie Delta turn out like those we got recently from LBJ country in West Texas, then we'll know for sure that rocks, like people, whatever their origin, are pretty much alike, once you wear down the rough edges.

Living it up at Las Vegas

W. B. Bleakley **October 22, 1973**

For more than 15 years I have felt like low man on the totem pole here at the Journal. Not that my name is on the bottom of the masthead or anything like that. It has to do with business travel.

George Weber and G. L. Farrar have covered the World Petroleum Congress in Frankfurt, Mexico City, and Moscow. John O'Donnell has to go to the pipeline contractors convention in Florida every year. Leo Aalund is flown to Puerto Rico to help dedicate a new refinery. Howard Wilson goes on a junket to Prudhoe Bay (preceded by Ed McGhee a couple of months earlier). And, at the time of writing, Bill Bachman is cruising on the Queen Elizabeth II to dedicate a refinery in Newfoundland.

Where do I go? Well, I hit the big oil centers, of course; Los Angeles, Denver, Houston, New Orleans, Dallas. But I also get some stories in real out-of-the-way places.

I once had an appointment to meet an engineer in Mary's Cafe in Red Oak, Okla. Then I covered two waterfloods in Kansas—one at Lyons, the other at Ulysses. And I can't forget the big well-servicing rig I studied at Coyanosa, Tex., or the gas-injection project near Duson, La.

But I had my chance this year. One of my usual assignments is to cover the annual meeting of the Society of Petroleum Engineers of AIME. The meeting site is historically New Orleans, Houston, Denver, or another of the cities on my usual beat. Now these meetings are always interesting and informative, and I always see many old friends and contacts. But I still felt I wasn't getting in on some of the goodies my colleagues here at the Journal seemed to get.

Then this year, SPE scheduled its meeting in Las Vegas! Now, I thought, I can join the other editors in the coffee room and talk about those far-away exciting places.

But do you know what happened? The meeting was a technical success. All the old friends and contacts were there all right, but they paid strict attention to business. The only time I saw them was during the technical sessions. I sat through several of these where speakers didn't have prepared texts, so I had to take notes. Then there was a story to write to meet the Journal deadline. Not counting exhibits of new equipment to study.

I stayed so busy I still don't know whether people have fun in Las Vegas.

Getting back to the office I had to withstand all the jibes and catcalls of the desk-bound editors who thought I was in Vegas living it up.

Come to think of it, I used to do the same things to those who went to Moscow, Mexico City, Puerto Rico, etc. Now I'm beginning to wonder if they had as little time for fun and games in those "oil centers" as I had in Las Vegas.

Snow on my boots

Jack A. Davis **November 12, 1973**

The first thing most people ask when you say you've been there is: "What's it like?" You don't get that kind of response when you return from Paris, Dusseldorf, or Beirut. But when you come back from Moscow, you have obviously been somewhere different. And believe me, it is different. About 1,500 Americans who also went to Moscow will agree with that.

The occasion was the first-ever oil and gas show put on in Russia by American oil-field equipment and service firms. A total of 189 firms participated in the Nefta-Gaz Expo '73 held Oct. 23-31 in Sokolniki Park, the Soviet capital's exposition center. Cost of the show to exhibiting firms was about $11.5 million, with nearly $30 million of U.S.-produced equipment on display.

The Oil & Gas Journal was there, too, with its own booth—the only publishing company so represented.

Doing booth duty in Moscow was like no other oil show we've ever worked or visited. The traffic through our booth had to be seen to be believed. They came from everywhere, men and women, young and old, Russian and Ukrainian, Polish and Hungarian, and others we simply couldn't identify. Other exhibitors reported the same Russian hunger for technical information.

There was brain picking on an impressive scale. The opportunity to examine and study the equipment and technological know-how of the finest oil-field equipment companies in the world was a unique event for the thousands of Russians who work, sheltered from the oil world outside, in the many oil and gas ministries and research institutes.

Russian attendance at the 9-day event topped 40,000. Hordes of fur hats thronged the avenues between the booths. Men visitors outnumbered the women as you might expect—but not by as much as you might expect. It's quite an experience to be confronted in the booth by a 50-year-old "housewife" with her shopping bag and to be asked, through our Russian interpreter, what we have published on the problems of transporting CO_2.

Get over that and a small party of middle-aged moms looking as if they are on their way to Gum (nearest thing in Moscow to a supermarket) come up and ask for a list of polymers.

Some of the exhibitors took giveaways to Moscow—ball-point pens, lapel badges, balloons, etc. These were a wow. I saw a surging throng of Russians clamoring around one booth for free books of matches. The exhibitor couldn't hand them out fast enough. One company took along a consignment of chewing gum to give away but had it impounded by Russian customs—that western habit isn't allowed in the U.S.S.R. Detente doesn't extend that far.

You come back from Moscow with other impressions, too. The grim fascination of Red Square, viewed in a bone-chilling wind from Siberia right at the time when the U.S. military was put on alert status. And not far away the delicate, magical beauty of the Bolshoi Ballet. To complete the picture, it even snowed.

Hardships in style

W. A. (Bill) Bachman **November 26, 1973**

On occasion, Journal editors are called on to endure some hardships in wilderness areas. We've had men on the North Slope in bitter winter looking at new wells. A few at new facilities in sweltering deserts. Others at uncounted new plants and fields in the boondocks. But all usually turn out to be humdrum assignments marked by some boredom and discomfort.

Now, John Shaheen has changed this. The party this New York industrialist staged for his ribbon cutting at a new refinery at Come by Chance, Newfoundland, surely must set a high for avoiding hardships and boredom. I can testify I suffered neither on his plant-dedication junket.

Shaheen chartered the swank liner Queen Elizabeth II to take 1,200 guests 1,100 miles by sea from New York to the bleak refinery site on Placentia Bay. There, the party was joined on shore by 3,000 Newfoundlanders who braved a chilly gale to help Shaheen christen his shiny plant and enjoy a feast on the grounds. The cruise then put back to sea and next evening sailed into Halifax, N.S., where provincial dignitaries and leading citizens came aboard for an evening of dining and entertainment. After another day of cruising south, the QE2 finally returned to New York.

Hardships aboard ship? Well, a fellow had to arrange his schedule by the clock or he'd miss a meal. There was earlybird coffee at 7 a.m. Breakfast (five elaborate courses if you wanted) from 8 until 10:30 a.m. Hot bouillon at 11 a.m. Cocktails from 12 noon until 1 p.m. Lunch (a gourmet menu) from 1 until 2:30 p.m. Tea with appropriate cookies and cakes at 4 p.m. Cocktails at 6:30 p.m. Dinner (a different menu of goodies and wine) from 8 until 9:30 p.m. And for those who couldn't bear the thought of retiring on an empty stomach, there was a midnight buffet of all manner of goodies.

Boredom? Well, in between meals the QE2 offered such amenities as swimming, Turkish baths, saunas, visiting tax-free shops, bridge tournaments, health classes, handicraft, skeet shooting, table tennis, afternoon and evening cinema, and a well-stocked library. In the evening, there was entertainment in the vast Queens Room and later in two cozy nightclubs. If these held little interest, the six bars opened at 11 a.m. and held forth until the wee hours, and the gambling casino was available at night.

Discomfort? The nearest to this was the formal dress required each evening. After all, this was a small price for such hospitality, since the gracious host was picking up the tab.

And speaking of the tab, as nearly everyone aboard did, it will quickly add up to more than $1 million. But Shaheen thought it money well spent. By the cruise, he was able to show international bankers and oilmen that a refinery could be built in the wilderness and two more were feasible. He soothed the feathers of any disturbed provincial politicians and business interests who had misgivings about the plants. He repaid many past favors for hundreds who had helped him.

John Shaheen, in case you haven't guessed, is one of the "doers" of the world. The energy and tenacity he devotes to a project is awesome. And it is going to take scores of "doers" like him to see us through our energy troubles.

"I'm still here"

Alvaro Franco **January 14, 1974**

There I was, deep in the lush jungle near Santa Cruz, Bolivia. And right smack in the middle of a little clearing I found what I had been looking for—a giant abandoned drilling rig collecting dust and rust under the scorching sun and constant rain.

While on vacation, I had flown to South America to interpret for John Cassidy, Jr., a Tulsa independent producer. In Santa Cruz, I heard of a "ghost" rig that was supposed to be somewhere in the jungle. I had to try and find it—and I did.

I wandered around, gazing in perplexity at stacks of unused drill pipe, a bulky and obsolete draw works, and an amazing collection of odds and ends, ranging from brush-covered engines to rusty shale shakers. Suddenly I heard steps behind me. I turned and there, seemingly out of nowhere, appeared a man.

He is a Brazilian, rapidly entering the sunset of his life. At one time he was a trucker working for a Bolivian operator. One day he was told: "We are shutting down our Bolivian operation. You are to remain here, watching the rig, until further notice."

That was nearly 13 years ago.

Every month, from somewhere, he receives a check. And that's all he knows. Month after month, year after year, he has been waiting, zealously watching the slowly decaying rig.

He has no regrets, no complaints. "But," he says pensively, "I'm getting old, and I'd like to go back to Brazil to die."

He eyed my camera and then said suddenly, "Please take my picture, and, if you ever run into them, show it to them. And tell them I'm still here."

Later on, in Brazil, I did try to track down the mysterious operator, but found no trace of him.

Meanwhile, the old Brazilian is still out there. Waiting for his next assignment.

North Slope in 2000 A. D.

Howard M. Wilson **March 18, 1974**

Did you know that in the year 2000 A.D. we will be moving 4,142,000 b/d of oil from the North Slope through two pipelines and by tanker?

By that year we will have produced a total of 26.6 billion bbl of oil and 22.15 trillion cu ft of gas from the slope.

But here's the kicker: Even after all this, the North Slope still will contain another 35 billion bbl of recoverable oil and 300 trillion cu ft of gas.

How do we know?

We learned it all from a forthright no-nonsense research paper which came to our attention while gathering information on some of the North Slope and trans-Alaska pipeline stories you will find elsewhere in this week's Journal.

The research paper was written by one of the nation's leading North Slope experts, John C. Reed, a geologist. Reed was once the U.S. Geological Survey's Man in Alaska and presently is a consultant to the prestigious Arctic Institute of North America in Washington, D.C.

Reed produced his paper in 1973 and called it "Oil and Gas Development in Arctic North America through 2000." He wrote it as though he were recording history in the year 2001.

His crystal ball says the trans-Alaska pipeline started flowing at 500,000 b/d in mid-1978, a year later than predictions made by the pipeline owners back in 1974. At the outset there was no environmental damage done by the pipeline. The big problems (Atlantic Richfield and BP Alaska, take note) were encountered in operating the oil fields.

But in 1981, according to this fiction, the line suffered a major break, about halfway between Bettles and the Yukon River.

"The break was caused by solifluction which destroyed about a quarter mile of line," our historian reported. "About 50,000 bbl of oil were lost." (Our dictionary doesn't carry the word "solifluction," but we trust the trans-Alaska builders will quickly find one that does.)

A gas pipeline from the North Slope and northern Canada was approved by Canada and the U.S. in 1974 and was operating by 1980. A few years later a second oil line was built, following the same international route.

Tankers began taking oil from the Canadian Arctic Islands through the Northwest Passage to the East Coast by 1980. It wasn't until 1995, however, that three 500,000-ton tankers began moving Prudhoe Bay crude by the same route to a new terminal off the Delaware coast. (Reed hedges at this point. If Offshore Nova Scotia, Newfoundland, and Labrador find big oil, scratch the Alaska tankers.)

The North Slope is booming in 2000. Barrow, a town of 3,000 in 1974, has 7,000 people and is a major center for health services, education, and research. A "thriving industrial center" of 3,000 has developed at Prudhoe Bay. While Prudhoe's male-to-female ratio is still high, the gap is closing year by year.

This is just a sampling. John Reed's history is labeled "Research Paper No. 62," dated September 1973. We don't know whether he has any extra copies, but we're putting ours in the Journal's "future file" for another look-see on Jan. 1, 2001.

Discovering oil

Leo R. Aalund **April 1, 1974**

Just a lot of people with unusual backgrounds have become experts on the oil industry this year.

All you have to do now is turn on the tube for the evening news or a special and some guy who you last saw describing retrofire on Apollo 13 is doing an "in depther" on the oil industry. It's "dominated," he's saying, "by 18 oil companies."

During the commercial when one of the three companies that dominates the auto industry is selling economy 10-passenger station wagons on this network, which is one of the three that dominates U.S. airways, you reflect:

It's rather nice, after laboring in the vineyard for a number of years as an energy writer, to find that the whole country is wrapped up in your chosen subject.

And though some of the material is still slanted, there are signs here and there that fact is creeping in.

You also think the Journal, though it is written for oilmen, may be playing a role in this.

Our subscription department is being deluged with requests for subscriptions. For months now, most of our editors have been inundated with calls from reporters and writers looking for sources of information.

For example, there has been a surprising amount of interest in this year's annual refining survey.

But you won't find it or details on it in any place else before publication. This editor knows which side his bread is margarined on. And just as important, we feel our cooperative industry respondents and subscribers should be the first to see it in print.

Of course, not all the newcomers to the oil business are in the media.

Several weeks ago a caller, who identified himself as a real-estate man, asked for advice on selling 40 million gal of Italian gasoline.

There would be a free lunch if the deal came off (boy, real magnanimous of him). But brokering is not our field and he got referred on.

Then, just yesterday the phone rang and the caller ebulliently said, "Hi, Leo, this is Moe with the Internal Revenue Service."

"I am filling it out," I said. "I thought I had until the 15th."

It seems though he wanted to talk about wax, or road-oil capacity, or something.

"Oh, that kind of IRS."

Journalistic technology—British style

Jack A. Davis **April 15, 1974**

The London staff of OGJ's sister publication, Petroleum International, met its own energy crisis recently by a fast-paced updating of the appropriate technology.

Electricity for lighting and heating was forbidden 3 days each week, but deadlines must still be met so auxiliary lighting was adopted—candles, actually.

The lumen output from one bare candle is inadequate for normal working so our research people launched a crash program to meet the situation head-on. An improved version was soon produced for mounting on a typewriter. Unfortunately, the candle developed a tendency to fall off into the lap of the user, which slowed production and kept the editors on edge.

It was decided to abandon that model altogether and concentrate on something more functional. A piece of circular cardboard was fitted into the metal lid of a round tube, and a hole for the candle carefully positioned to discourage the flame from catching the reflector on fire too often. The reflector itself was a piece of copy paper curved around the rear half of the lid and sticky-taped into position. It was judged to be safe under normal conditions and besides, there was always more copy paper. This new, improved version was called the Mark II.

But the frenzy of inventiveness had not played out. If one candle is good, two must be better. Careful experimentation revealed that another hole, judiciously angled, could be located in front of the first one, and we were soon the proud possessor of a two-holer.

Operators report that both the Mark I and the Mark II versions have completed many hours of relatively trouble-free service and normal office output has been maintained at around 750 cwph (candle words per hour).

One lighting unit using six candles was also developed for special projects but the heat generated outweighed the lumen advantages and we didn't want to burn the place down.

We feel that we've led the industry in this research, and we are thinking of presenting a technical paper on it.

Still a bargain

Jim West **April 29, 1974**

Oilmen have told us for a number of years that the price of energy was going to have to go up.

And sure 'nuff it did.

Top prices for U.S. crude, for instance, rose to \$10/bbl or so last year, although most domestic crude carries a price-controlled ceiling of \$5.25. New-contract prices for natural gas have risen to almost \$1/Mcf in some areas during the past couple of years, but area rates for interstate sales keep hovering around 19-23¢.

With the increased prices—for crude in particular—came reports from the general news media that tended to show the prices are too high and should be rolled back.

That's just not so, says Frank Ittner of Midland, Tex. He's a former exploration vice-president of Superior Oil Co. and has been an independent geologist since 1959.

Ittner is convinced that crude is still a good buy, and so is natural gas.

To prove it, he roamed the aisles of a local discount store jotting down the prices of several items, some of which were displayed at special sale prices. Then he converted these figures to a price per gallon and compared them with crude oil and natural gas.

Here is his list, passed along to us by Warren Hammel, a Midland scout for Superior:

Commodity, \$/gal
Old Confidence, 30.00
  100 proof
Honey, 11.00
Baby lotion, 7.92
Linseed oil, 6.00
Corn oil, 5.80
Turpentine, 2.99
Beer, 2.50

Commodity, \$/gal
Apple juice, 2.10
Orange juice, 1.85
Milk, 1.65
Coca Cola, 1.46
Vinegar, 1.43
Tomato juice, 1.28
Clorox, 0.58
Distilled water, 0.49

By comparison, crude oil priced at \$5.25/bbl costs 12¢/gal, and natural gas at an area rate of 19¢/Mcf costs 3¢/gal with heating value converted to crude equivalent.

The Midland Reporter-Telegram published the list and comparison, but so far it's the only newspaper to do so.

"Should tomato juice be worth 10 times a gallon of crude oil?" Ittner asked in an interview. "Should distilled water be worth more than three times as much? Probably not.

"In fact, a good case could be made that crude has been a great bargain since the East Texas field was brought in. And at current prices, it still is."

Learning to trust your fellow man

Ed McGhee **May 6, 1974**

An editor has to trust people. He can't be eyewitness to every momentous event that occurs. He can't caliper every joint of pipe nor read every pressure gauge. So he judges people to decide who tells it like it is.

Usually it's no problem. But here's a case that troubled me. Red Carstairs is drilling foreman for Panarctic Petroleum Ltd., and there came a time when I wondered if he was pulling my leg as far as a human limb will stretch about a Far North drilling project.

We stood looking up at a drilling rig, and Red began telling me that we were 8 miles out in the ocean, 400 ft above the bottom of the sea, and there were 8 ft of ice frozen beneath us.

He did show me a hole filled with green water, but when I jumped up and down on the ice around it, it felt like it was on terra firma to me. "It dern well better feel like that," Red agreed, "or it couldn't support that rig."

Now about that rig. There wasn't a barge nor an anchor, nor anything marine in sight. Not even a Cajun roughneck. Anyone could see that it was an ordinary skid-mounted land rig. We had climbed down under the derrick floor and Red pointed out the water standing in the cellar. (He insisted on calling it a "moon pool" like they have on regular drillships.)

As I took his photo, he was telling me that a seal surfaces regularly at this spot to breathe. Something told me to stick around and see for myself but since it was minus 40 degrees F. (I didn't need a thermometer to verify that), I decided I'd take Red's word about the seal.

Finally I decided that if it was all a hoax, it was a most carefully contrived one. And so, it all must be true about drilling through the ice. The Editor could hardly fire me if someone of Panarctic's prestige had gone to such elaborate lengths to mislead me. So when I got back home I duly filed the story which appeared on p. 23 of the Apr. 15 Journal. "Panarctic pioneers floating-ice drilling," the headline says (there's also a story beginning on p. 132 of this issue).

So far, no incredulous reader has written in to tell me I've been duped. But if someone does expose a fraud, remember this: I didn't claim to have seen the seal.

It ain't always "finders, keepers"!

Frank J. Gardner **June 3, 1974**

As any reader of the Journal must know, this reporter has always been a North Sea bull. From the very first discovery at Ekofisk, we've insisted that when you find one giant oil field somewhere, that isn't the end of it. You're bound to find more.

Although we're rockhounds, our faith in big oil for the North Sea doesn't stem from any extensive knowledge of North Sea geology. Rather, from 35 years of watching oilmen, writing about them, following their tracks across the oil map, and adding green and red spots in their wake.

We used to enjoy writing happy news about those oilmen, celebrating their discoveries, admiring their abilities, wishing we were they.

But these days, happy news about the oil industry is hard to come by—except in the North Sea, where we've been ecstatically adding green and red spots to our North Sea oil and gas map for the past 9 years.

But, as with any successful project, the politicians have spotted the North Sea—and that's bad news.

We had hoped the pattern in the North Sea would be a little different from the one that's developed in so many oil-producing states. "We have no oil. Please come and look. We'll give you a free hand, low royalties, low taxes, low rentals, no bonus, anything to find us some oil."

So the industry moves in. And it finds oil. Big oil.

So, up go the royalties. Up go the taxes. It's a sin to make a profit. Back in 1970, in pointing to the advantages of the new North Sea oil fields, we referred to them as "this politically safe source of supply."

We take it all back. It ain't all that safe.

Norway hasn't nationalized, but it's certainly come close, with demands of 80% participation.

Britain's Labor party screams for nationalization. And the Tories, before they goofed, intended to put the squeeze on the oil companies.

The Dutch have been more lenient, with 40% participation options.

The Danes have been quiet. After all, they don't have big oil yet.

The Germans haven't said a word. They have nothing at all ... yet.

So you see, the amount of political dabbling is directly proportional to the amount of oil found.

It reminds us of an incident in Oslo a couple of years ago. We were interviewing a high government oil official. "Why," we asked, "did you set royalty payments so low?" They were 10% at the time.

His answer? "We didn't think they'd find a damn thing!"

Career opportunities revisited

Ed McGhee **October 14, 1974**

You remember that big colorful section in the Sept. 16 issue. It was called Career Opportunities in Oil. In it, Bob Lair and a couple of other Journal editors were making a case for the industry as a good place for young people to dedicate their working lives.

Only problem was that they put all their material together while I was out of town on a trip. If they'd waited a couple of weeks, I could have added a jim-dandy career opportunity to their list.

The career I'm referring to is that of a man I ran into while I was on that trip. He works for National Iranian Oil Co. on Kharg Island. He's a gardener. He plays an important role in NIOC's effort to run an efficient, low-cost operation at the world's largest crude-export terminal.

Kharg Island, as you'll remember, is a bleak, arid bit of ground barely protruding above the surface of the Persian Gulf. It can't generate enough agricultural produce to feed a tiny fraction of the 1,000 persons who are directly employed in the terminal—much less the 9,000 other persons (families of workers, mostly) who live on Kharg.

Kharg offers few enough amenities without one's having to subsist on a diet lacking in greenery. So, NIOC flies in fresh vegetables and fruit daily. The cost, naturally, is high.

That's why NIOC is testing what is called "environmentally controlled horticulture." Aim is to cut the air-freight bill by raising eggplant and squash right there on Kharg.

Without getting into how NIOC came into this particular bit of experimentation, it is now operating two very productive acres. Using a system developed by the University of Arizona, it can on those 2 acres grow all the tomatoes Kharg residents (and visitors like me) could consume.

Environmental farming requires only seawater (which Kharg almost drowns in), sand (which Kharg's beaches supply in abundance), chemical nutrients (brought in by barge), and a little fresh water (piped over from NIOC's seawater-desalination plant).

Evaporation of the seawater cools and humidifies the "greenhouse." Inside that translucent-plastic-covered structure, the vegetables are rooted in the sand and fed by the chemicals dissolved in the fresh water. It's everything your average cucumber vine could wish. Result: big, healthy vegetables in astonishingly short time and in astonishingly little space.

NIOC doesn't feel it has all the answers yet. But if the early indications prove out, you may be seeing help-wanted ads for horticulturists alongside those for tool pushers.

The API and public relations

Robert J. Enright **November 11, 1974**

This year, the American Petroleum Institute tries still another tack to improve oil's incredibly bad standing with the public, press, and government.

It is pulling 34 affiliated state groups back into the fold—they once were part of API but were separated in 1958—to more directly supervise and hopefully improve effectiveness of public and government relations.

We hope it helps. If it doesn't, the industry is in for it. It's difficult to appreciate how bad the public's concept of the industry is—or understand how it got that way.

How bad is it? Prof. David Gottlieb, acting dean of the University of Houston College of Social Sciences, can tell you. Gottlieb surveyed the heads of 900 Texas households—Texans, mind you—during and after the Arab oil boycott. He found this:

- They believe there was a temporary energy shortage—but no long-term crisis.
- They believe the shortage was precipitated by the oil industry and that when fuel prices moved high enough and environmental controls were eased fuel would be plentiful.
- The best-educated and the young (and this is alarming) are the most skeptical, believing that the energy shortage is contrived and doubting the need for conservation.
- Most of these Texans formed their opinions on the energy situation based largely on network television newscasts and programs and believe national TV to be the most reliable source of such information!
- The conviction is widely held that oil's television ads are cute—but that they don't tell the truth.

The oil industry is suspect. And little wonder in view of the spate of "sensational disclosure" type stories which appeared so profusely in the news media this past year.

API Chairman Charles Spahr bemoaned this in Dallas a few weeks ago. Dozens of oil "conspiracies"—all untrue—were bannered all over the country. The public was advised that hundreds of tankers were lying offshore waiting for prices to rise before unloading; that thousands of wells were capped in the Gulf of Mexico waiting for prices to rise before producers would produce them; that industry statistics were deliberately rigged to disguise the true supply and storage picture; that American oil companies were exporting huge quantities of oil while American motorists waited in lines for gasoline; that the Arab oil cutoff hadn't reduced oil imports.

Each of these charges was made, repeated, and made again. Each was investigated thoroughly and found false. Yet, Spahr points out, "almost 9 months after they were proved wrong, we still find reference to them in the nation's press."

Is it any wonder then, he says, that the public believes the shortages were the oil companies' fault? No, it isn't. And the API has a monumental job.

They're going to raze the Rice

Leo R. Aalund **December 16, 1974**

Even before the news made the headlines, most of downtown Houston had heard the sad story by word of mouth.

It was simple: The Rice Hotel will be closed on Mar. 31, torn down, and the land sold.

Generally, the closure of a hotel is a local story and only rarely rates a headline in a metropolitan paper. But the Rice is different.

It's probable that more oil history has been made in its rooms, bars, and restaurants than any single place in this country. My colleague, Gulf Coast Editor Dan McNabb, doesn't totally agree. He thinks the long-gone Tulsa Hotel is a strong contender for this distinction.

But that was primarily exploration and production country, and anyway Dan is not yet 30, or over the hill, as he puts it.

Legions of wildcatters have checked into the Rice, too, for fun or business. But it has also housed many of the wheeler dealers, planners, and visionaries who started putting together the great petrochemical and refining complex on the Texas coast, not to mention scores of oil meetings and technical sessions. So this editor thought that many in the industry around the country would want to hear of this story, which is heavy with nostalgia.

Oil is only part of it.

The corner of Texas Ave. and Main, where the Rice stands, was the location of the first capitol of the fledgling Republic of Texas.

The area is steeped in Texas history and has had an inn or hostelry since the days of the republic.

The present Rice, still the biggest hotel in Houston, was built (of real bricks) in 1913.

In 1928 it was the headquarters for the Democratic National Convention where FDR nominated Al Smith to carry the donkey party's presidential banner.

And there in November 1963, John F. Kennedy ate his last evening meal before starting that tragic trip to Dallas.

Ironically, the hotel is named in honor of William Marsh Rice, who bequeathed the money to found Houston's famed Rice University, which now owns the hotel. The school's board of governors says the hotel has been losing money of late and doesn't think it should stick university funds into a modernization program.

Rice University got the hotel as a gift in 1971 from Houston Endowment, which was established by Jesse Jones, a nationally known figure in the Roosevelt era and one of Houston's mightiest movers. Jones died in 1956.

The first real competition for the Rice came in the early 1950's after Wildcatter Glenn McCarthy built the Shamrock "way out" on South Main. Conrad Hilton now has the action there.

Motels and airports also took the edge off the Rice's prominence in following years. But it never went to the dogs. Oilman John Mecom almost bought it in 1966 along with the Houston Chronicle, another part of the Jones empire. But the deal fell through at the last moment.

But it looks bleak for the Rice now.

Things have changed in 37 years

Frank J. Gardner **December 30, 1974**

Just recently, as we were beginning to compile the statistics for this annual worldwide issue of the Journal, a friend of ours—Pres. Richard Davoust of Quasar Inc., Evansville, Ind.—sent us a copy of a most interesting map, along with a nice letter saying he'd found it while cleaning out some files.

It's a big wall map showing oil and gas fields of the world which was inserted into the Worldwide Issue of the Journal in 1937!

In short, it's mostly white space. We never realized how much things have changed in 37 years (we were still in college then but avidly following the oil business and reading the Journal in the school library).

Would you believe not a single offshore oil or gas field?

Since then, the industry has discovered about 150 billion bbl of oil and 80 trillion cu ft of gas in offshore fields. Well, there is one green dot in the middle of Lake Maracaibo, but we usually don't count lakes as truly offshore, since we've come to think of the word in terms of continental shelves.

According to the United States Geological Survey, offshore exploration is active off more than 80 countries today.

And speaking of countries . . . there's been quite a lot of change there, too! The colonial era was still with us in 1937 (and it just ended this year), and a lot of countries on today's oil map weren't separate entities then.

For instance, there's no Algeria on the African part of the map. A huge area is denoted "French West Africa," another "French Equatorial Africa," and on the east is "Italian East Africa."

The only fields shown on the African Continent are a couple of red gas dots in Algeria, one in Morocco, and in Egypt are two green and three red dots. What, no Libya? No Nigeria? Angola? Gabon?

Switching our gaze to the Middle East, some strange things are noted. Not a single oil field in "Arabia" (no "Saudi" prefix!). Only Bahrain is shown with a field on the southern coast of the Persian Gulf. To the north, three fields are shown in Iran— Naft Kanah, Haft Kel, and Masjid-i-Suleiman.

The North Sea, of course, is blank.

Studying this old map, we can't help but wonder. Will the maps we're using today in our offices be as amusing to us in 2011 as this 37-year-old relic is now?

If we know our oil industry, the answer is "Yes, fool." That 1974 map doesn't show the Antarctic fields, nor those in the Barents Sea. Why, it doesn't even show the mid-Atlantic fields! What kind of oil finders were they back in '74? That was before the magic black box, of course!

Instant oil experts

Larry Auldridge **January l3, 1975**

The telephone buzzer in our Journal office in New York has a comic sound—something like Donald Duck being throttled by Bugs Bunny. And we hear that buzzer often.

Most of the time we're answering calls from responsible persons or organizations who want—and get—what oil and gas industry information we have available.

But, there is a new category of callers now proliferating on the New York scene. These are the non-industry "instant oil experts."

One such caller informs us he is writing a free-lance article on the oil industry, which he hopes to sell. He wants, he says, "a few pointers to make me sound at least a little authoritative on the subject."

Another caller announces he has just been appointed energy editor for a news organization and "would like a little fill-in, since I honestly don't know much about the oil business."

And still another person who identifies himself as a news reporter wants "to get some dope on oil companies that caused the oil crisis."

A young lady working for her boss down on Wall Street asks, "Would you please give me the titles and dates of all the stories you have printed on the North Sea?" Asked if she has checked the library of her very large firm, she admits she hasn't but says, "I thought it would just be easier for me to call you."

Now, Donald Duck nearly gets the works when a fellow calls to say he's preparing a table on world crude-oil production and he would like numbers for the last 12 months. We quickly pull from our bookcase a binder with up-to-date figures on the subject and read off the numbers for the totals. No, he caller says, he'd like to have a month-by-month, country-by-country rundown. In declining, we tell him this amounts to as many as 1,000 numbers. But persistently, the caller replies, "That's all right, I have the time."

These are just a few of what we refer to as our "Donald Duck" calls.

But more seriously, this editor feels that he and his co-workers on the Journal —and industry representatives as well—have a responsibility to try to improve and expand public knowledge and understanding of the industry.

Sincere oil-industry students always merit all the help they can get to make factual, unbiased presentations concerning our industry. On the other hand, we must try our best to serve the untutored, self-appointed oil-industry mentors. It's not impossible. But isn't it frustrating?

Economic motivation in Tehran

Ed McGhee **January 20, 1975**

Ever since 1908 when oil was first discovered in Iran, a large mutual-admiration society has grown up between Iranians and oilmen.

Oilmen from abroad have always liked living in Iran. In turn, the people there have always responded with hospitality to foreign drillers, pipeliners, etc.

It didn't stop with simple good relations between two unrelated groups. The Iranians themselves took naturally to oilfield work. Today men from Isfahan, Abadan, and Shiraz occupy responsible positions in engineering offices, oil fields, and plants all over the world.

On a recent trip to Tehran, I asked an American friend who'd lived there for many years if he could explain the Iranian phenomenon. Other nations seem to have the same opportunities but make less of them. It was his opinion that Iranians are just naturally industrious and—he added meaningfully—they can be "economically motivated."

I didn't realize, but I was about to get a lesson in economic motivation.

When I left the man's office, I began the short walk back to my hotel. As I reached a street corner, I was approached by a lad of perhaps 8 or 9 years who wore a most engaging smile and carried a shoe-shine kit. He spoke only Farsi and I only English, but there seemed no question about the nature of the business at hand. So I instructed him to proceed.

He set to work at an industrious pace, smiling and talking brightly in Farsi—remarking about the weather, I supposed. He quickly indicated that it would all go better if I removed my shoes and donned a pair of slippers he'd brought along for just such a purpose. This was a new procedure to me, but not an unreasonable one I thought.

The next thing he did was also new to me—he scraped the soles of my shoes. Not a bad idea, I thought.

As he proceeded with his task, I glanced away to contemplate one of those traffic snarls that have become a 24 hr/day feature of Tehran streets. When I glanced back, my young friend presented me with a lustrously shined pair of shoes—which also sported new half soles. Unneeded half soles, of course.

This was hardly the first time I've ever been "had." But it was the first time I'd been taken by someone so young. The $2 it cost me was cheap price to pay, I thought, for a lesson in economic motivation.

California and potato salad

Howard M. Wilson **February 17, 1975**

After 5 years on the California beat, we should be fully prepared for the kind of unreasoning arguments which are habitually thrown into the debates on offshore drilling.

This year we had fervent hopes for a change.

After all, just a year or so ago our friends in the L. A. Basin were soberly discussing the possibility that they, along with 10 million other Angelenos, would have to abandon their paradise on the Pacific. With no gasoline, what was to propel them 50 miles each day along the freeways to work and home again? The Arabs had shut off their oil.

And during those days, the producing countries tripled the price of crude. Gasoline at the pump has climbed into the 50-60¢ bracket.

The price of electricity and natural gas also zoomed.

We live with the constant threat of another embargo that could cripple us. What's more, this threat is one that won't go away. It will be nudging us for years to come, perhaps a generation.

The only escape, barring the arrival of the millennium (which some apparently expect), is development of our own energy resources.

The other day we went out to the Interior Department hearings in Beverly Hills to observe whether any of these realities are any better understood by southern Californians today than they were, say, 16 months ago.

We were not encouraged by what we saw and heard. True, the antioil crowd wasn't as large, or as raucous, as in other times. But its political clout has grown. The city governments of southern California have stepped into the leadership role in opposing offshore leasing.

In their zeal to prevent "unsightly" drilling rigs and platforms from dotting the seascape, they trotted out every shop-worn argument they could lay hands on to delay or kill future offshore operations.

Their chief contention was that the United States needs a formalized national energy policy first. Drilling for offshore oil can come later.

As they kept repeating this pitch, we had the vision of a patient dying of peritonitis in a hospital bed. Gathered about him were several surgeons discussing the possibility that medical science might develop a method for curing his ailment without drawing blood.

If we may pursue this point with a mixed analogy, consider the problem of the potato-salad committee for the church picnic. No one had told the committee how many picnickers to prepare for. So they decided not to have any potato salad.

Who wants potato salad at a picnic anyway?

We believe the committee ought to start peeling potatoes.

Arabs know how to treat the press

Frank Gardner **March 31, 1975**

We just returned from attending our first Arab Petroleum Congress—the ninth such gathering—held in Dubai.

"Dubai? Wherezatt?" was the usual raised-eyebrow response from friends to whom we mentioned the trip. We know where it's at, of course, because it's become a substantial oil producer over the past few years.

We'll say this—the Arabs really know how to put on a show. The congress was well-organized, well-run, and well-attended—over 600 people, mostly Arab but with a substantial non-Arab attendance.

It was quite a scholarly meeting, too. The papers covered everything from oolites to wellsites. And another good thing—every paper (a couple of exceptions) was printed and available before the congress opened. We know, for we carried a complete set up in the elevator and barely made it to the room. They weighed a good 25 lb.

Speaking of the room (the meeting was held in Dubai's spanking new Intercontinental Hotel, specially opened for the congress), it wasn't quite ready for us. But after the first day or so, the phones began to work, the doors began to lock, and the water started running.

Star of the show was Mana Saeed Otaiba, oil minister of Abu Dhabi. Since

Dubai has no oil minister per se, he acted as host minister on behalf of the United Arab Emirates, which includes both Arab countries.

Otaiba is an impressive speaker. Using no notes or text, he delivered his opening-session address forcefully and with feeling.

On one occasion we really felt for him. On the fourth day, he held a press conference in a comparatively small room that was literally packed with reporters, television crews, and photographers. We counted 12 tape recorders on the table before him, and other microphones and cameras were suspended over his head.

For the better part of 2 hr, the young (29) minister endured the heat of the lights and of an Arabian-Gulf sun that shone in his face. But he is extremely courteous to the press, and no question goes unanswered. In fact, his answers are minispeeches.

We saw him on television a bit later, and he's just as impressive on the screen as in person.

Speaking of that, it was our first time to see color TV in an Arab land. There are all-Arabic programs that were completely unintelligible to us, but there are American programs with Arabic subtitles as well. We didn't see "All in the Family," but there was a great Martha Ray comedy!

All in all, a most pleasant experience in Dubai. But on our departure, we had the most pleasant surprise of all. Checking out at the cashier's desk, we were told: "Oh, there's no charge for your room. You're a member of the press!"

Mr. Hilton, are you listening?

Offshore 1984

Dan McNabb **April 28, 1975**

ACT 1. The scene is the Gulf of Mexico, 30 miles east of the Mississippi Delta's South Pass. Time is 9:40 p.m., Jan. 16, 1974.

The derrick barge Foster Parker guides a 462-ton lower deck onto a jacket standing in 340 ft of water. About 30 hr later, at 3:40 a.m., Jan. 18, the second deck goes on.

Everyone is happy and relieved—except a disappointed photographer on the crane barge. He had planned to record the events in living color but insufficient lighting in the morning darkness forces him to shoot the scene in black and white.

Now for Act 2. The scene is the Atlantic waters off the Outer Continental Shelf. Time is nearly 10 years later, past sunset on Oct. 27, 1984.

Workmen are installing the first drilling and production platform on the

U.S. Atlantic coast. Jacket piles have been trimmed, and the tackle is ready for the deck lift. Word has come down, however, to wait until sunlight for the lift.

In a cramped cabin on the derrick barge, a bureaucrat from Federal Oil & Gas Co. (Fogco) explains the shutdown to an agent of the contractor, worried about the weather and wanting to set the decks as soon as possible.

"The President wants to be here for the lift and conduct a ceremony," says the Fogco bureaucrat. "The lift must be made on a sunny day so the TV people can get good shots."

"Why this spot?" the contractor's man wonders aloud. "No wells have been drilled within miles. There're many more likely drilling sites."

"There's at least a slim chance of finding oil and gas," retorts Fogco's man. "Besides, the President ordered this platform himself. He wants an offshore platform for his pre-election speech on nationwide TV. He wants to show the voters he's really cracking on the energy shortage."

Act I is for real. And Act II may be too, if Congress adopts the scenario now being proposed.

Freshmen congressmen go to head of class

Caroline E. Mayer **May 12, 1975**

Freshman everywhere used to always be easy to identify—by their timidity, naiveté, and deference to their elders.

But not so with the freshman class of the 94th Congress, which doesn't believe that older necessarily means wiser. The freshmen have destroyed the stereotype of the first-termer who by tradition is supposed to be seen and not heard.

The 44 new Democrats in the House couldn't be more different. They have marched right to the head of the class, not hesitating to speak out on any issue, push legislation (sometimes successfully), and take on the established big men of Congress.

On occasion, they have become its muscle, particularly in the Democratic caucus, where they were influential in deposing three committee chairmen and repealing percentage depletion.

But the freshmen have stubbed their toes from time to time, particularly with their energy program. Thirty-nine Democratic freshmen endorsed a package that called for a price rollback in domestic oil prices to $4.25/bbl for old oil and $7/bbl for new oil; no increase in gasoline taxes; a mandatory national gasoline-allocation system to reduce gasoline consumption by 5-8%

the first year; and mandatory standards and tax incentives to enforce energy efficiency for automobiles, appliances, and residential and commercial construction.

Congressional observers were dismayed, stunned, and disappointed to find the promising new members behaving like old politicians. Reporters, lobbyists, and committee aides, alike, repeatedly criticized them, muttering, "They all want something for nothing."

But when they appear to fail, the freshmen are not without influence in shaping legislation.

In the ways and means committee, for example, they played a key role in toning down Chairman Al Ullman's (D-Ore.) original 5-year, 40 cents/gal gasoline tax. The latest proposal is for a 3 cents/gal tax in 1976, with possibly as much as 20 cents/gal more tax being imposed in future years, depending on whether the rate of gasoline consumption rises.

But this tax, and other key provisions of the energy tax bill, face a further scaling down as the new Democrats seek to revise the bill when it gets to the floor.

Timid? Naive? Unfailingly polite to their elders? Not these freshmen.

Perhaps one reason is that many were political veterans, either elected officials or activists with experience, before coming to Washington.

As one congressional aide says, "They may not get all that they want, but they could block anything sensible from going through Congress."

Most Congressmen don't learn this until their second term.

It is written...

Robert G. Lair **June 16, 1975**

> **style (stil)** the custom followed (as in a business, editorial, or printing office) in spelling, capitalization, punctuation, and typographic arrangement and display.

That's the way Webster defines what Journal editors try to follow in preparing each week's issue.

The first thing a new editor is handed—perhaps even before his W-4 form—is a big yellow 23-page notebook labeled "Style Book" with the briefest of instructions: "Learn it, chapter and verse!"

And with good reason. Journal editors have discovered over the years that a well-chosen consistent style aids in avoiding errors, facilitates reading for our busy subscribers, and improves the magazine's appearance. A publication that sometimes spells out a number in one situation and uses it in figures in a like situation is increasing the possibility for mistakes. A reader in a hurry having to stop to unravel a style confusion may stop reading the publication permanently.

Although our strict adherence to style in preparing each week's Journal turns out an easier-to-read product, Journal editors often find it's not an easy task to rationalize some of the more intricate style rules. The nuances of some rules have brought many an editor to his knees begging for mercy.

For example, not a day ends that some style-beleaguered editor—seeking a nice, neat black-and-white answer—doesn't ask, "What do we do with government?" Nowadays, a typical oilman would have a ready retort for that one. But the puzzled editor simply wants to know whether to put the word "government" up or down—that is, capitalized or not capitalized.

The office style "expert" responds, "It all depends."

Beleaguered editor: "Depends on what?"

Expert: "Depends on how you're using it."

Editor (sarcastically): "As a word, in a sentence, in a paragraph..."

Expert: "I know, I know. But is it a noun or an adjective?

Are you talking about the U.S. Government (note the capital "G") or a foreign government? Or is it a state or provincial government? Does the word stand alone, or is it preceded by the country's name? These things matter, you know.

Editor (mumbling): "No. No, I didn't know."

It's at this point the frustrated editor usually hits the floor with his hands clasped and facing heavenward. Or the so-called style "expert"—sighing heavily—will make his way toward the editor's typewriter, scrutinize the offending word, and decide the proper usage. And the editor's sanity gradually returns—until next time.

Fortunately, most Journal style rules don't cause such headaches. And most editors have long since given up asking why or trying to find the logic behind some of the more complex rules—although it's always there, somewhere.

Experience has shown they work. They ease the editor's job—believe it or not—and, above all else, make reading the Journal a little smoother.

To the shores of Tripoli ...

Gerald L. Farrar **July 14, 1975**

The signs of prosperity are everywhere in Libya these days.

I flew into Tripoli in mid-June on an oil-related assignment and began to see the activity immediately. A new refinery-storage terminal installation looms near the airport. Then a little farther on, near the outskirts of the city, row after row of spanking new apartment complexes, some in use and some still under construction, rise in profusion.

Just like the major cities in other countries, Tripoli has a traffic problem. New cars from Japan, Europe, and the U.S. crowd the streets in all sections of the city, at all hours of the day.

The Libyan Palace Hotel is just a block from the busy harbor, and the sight at night from the seventh-floor roof garden is breathtaking. The lights of ships are visible as far as the eye can see, to the edge of the horizon in all directions.

These ships are waiting their turn to berth and unload at the Tripoli docks. It takes 3 months from the time a ship drops anchor and registers with port authorities until the dock-side stevedores swing into action to unload cargo.

Libya's two major ports— Tripoli and Benghazi—are simply not big enough to handle the crush of industrial equipment and consumer goods coming into the prosperous country. And the Libyan government is moving ahead with plans for a third major port to help relieve the congestion.

Most of Libya's income from outside the country comes from oil. So, a great deal of the equipment imported into the country is earmarked for oil development.

Libya's current production is about 1 million b/d, off somewhat from the peak levels of a few years ago. But the national oil company and the private companies operating in Libya are maintaining a high level of exploration and development activity. The country will be able to sustain a producing capacity well in excess of the present level for many years to come.

The new construction and other developments are not limited to Tripoli and Benghazi, either. Oilmen who travel throughout the country say it is everywhere. I can vouch for it as far as Sabratha, the site of an ancient Roman city some 50 miles west of Tripoli. New apartment houses and new public buildings are rising in every village and town along the way.

Public education is high on the list of priorities. Libyan children attend school, receive books and supplies, and have a hot lunch each day at no cost to them. Promising secondary-school graduates may take advantage of liberal scholarship aid in attending universities in Libya and abroad.

All in all, the government seems to be spending its oil money to educate and raise the standard of living of the Libyan people—a worthy objective by any standard.

In quest of Calinectus Sapidus

Leo R. Aalund **September 1, 1975**

He hurt me when I made my first thrust, and I was bleeding slightly. He now had the advantage and his hate-filled brown eyes said there would be no quarter given in this battle under the torrid afternoon sun. So I snatched the only weapon at hand, a paddle, knocked him on his back, swiftly grabbed him from behind, and threw him thrashing into the ice chest—a fine male specimen of a great blue crab, or Calinectus Sapidus (CS), with a claw span of some 10 in.

The cut finger would smart Monday morning when typing the final touches to our energy-conservation special report. But it was a trivial price to pay for another of these tasty crustaceans from the sea whose claws and bodies rival in appearance the world's finest blue and creamy-white porcelain.

But there is an easier way to snare CS than bringing him up over the side of a small skiff on a line baited with a ripe fish head. Easier, that is, if you can get to Coteau Holmes, La. This settlement, which is not on every map, is on the edge of the Atchafalaya basin a dozen or so miles from St. Martinville. But finding it is a pleasure in the lovely Bayou Teche country.

You wouldn't know you are in Coteau Holmes until you see rows of crab traps, stacked higher than a man, on a lawn. They are made of rubber-coated chicken wire folded into a box shape. The tricky part of the construction is a passage that permits CS to get to the bait, but then directs him and perhaps two dozen or so of his brethren into an escape-proof holding chamber.

Walk around a few rows of traps until you find a carport festooned with loaded (in season) cucumber vines. You're at Mrs. Guidry's place then. A knock on the back door will probably elicit a "Come in."

Be ready for chat in Mrs. Guidry's living room. She is an energetic lady who built most of her own house because her husband had to spend so much time on offshore rigs.

She taught herself to build crab traps, which involves some skillful handwork, then got her neighbors involved in a sort of backwoods assembly line. This, she says, brings a little cash into an area with very limited employment opportunities.

She sells the sturdy traps for $4.42 apiece, but is prepared to make bigger deals. If you want about 300 traps, she may even load up and meet you in Biloxi or Port Arthur. They work well!

The truth according to John Moss

Caroline Mayer **September 8, 1975**

"I have a lot of constituents," Rep. John Moss (D-Calif.) trumpeted recently in his hearings on the natural-gas industry. In fact, proclaimed Moss, "I have . . . a national constituency" as chairman of the House commerce subcommittee on over-sight and investigations.

So, erupted Moss, "I will be damned if I am going to sit here and see this record anything but complete and accurate, and we are going to get the facts. We are moving to get them from the producers themselves, and we will treat them fairly. But we are not going to be stampeded into any action of deregulating."

The volcanic Moss has a mission, one from which he and his subcommittee staff will not be deterred. He is out to show "the depth of avarice" in the petroleum industry and thereby stem the tide of sentiment building in favor of natural-gas deregulation. The forum for the campaign is his subcommittee; the weapon, the hearing record.

His mastery of parliamentary procedures, based on his 22 years in the House, will certainly help him shape the record to serve his purpose. The hearings to date demonstrate the special meaning of his pledge for accuracy and fairness.

For example, Moss has his chief counsel grill a friendly witness for nearly 90 min. Then, when it comes time for the Republican proponents of deregulation to cross-examine, the imperious chairman invokes the 5-min rule.

Moss also likes to keep the Republicans in the dark about hearings or staff reports until the last minute. His staff of 17 professionals and eight clerks sifts over the mountain of investigative data in secrecy for 3 months, then springs it on the one lawyer and secretary, who comprise the minority staff, a week before the hearing.

But Moss is unwilling to let even his carefully custom-tailored record speak for itself. During the hearings, it is not uncommon to see his staff buttonhole reporters from major metropolitan newspapers to regale them with horror stories of industry corruption.

He also likes to keep the public record neat, uncluttered by criticisms of a highly publicized staff report that charges a company with lying and withholding gas to get higher prices. When an opposing congressman issued his own press release denouncing the report, Moss's staff sprang into action. An aide hastily called reporters who received the critical release to assure them of its falsity—an action unprecedented in the experience of one veteran Texas reporter.

Yet Moss repeats he has a "responsibility" to make sure the record is "accurate" if not complete.

What he means, translates one Moss observer, is that "everything in the record must be accurate, not necessarily that it accurately reflects all points of view."

So insistent is he in purging all dissent that one can only wonder just how confident the congressman is in the ability of his case to stand on its merits.

Will we miss all the fun?

John L. Kennedy **September 15, 1975**

Reading the special drilling section in this issue may give the impression that all the fun is being taken out of drilling. Articles deal with, among other things, collecting information from the bottom of the hole and accurately predicting reservoir pressures. These articles indicate there may be a real danger that drilling has become a science rather than an art.

And as any child knows, science is dull and art is fun.

What made drilling "fun" in the past were the simple pleasures of fighting lost circulation; the excitement of not knowing what type of formation was being drilled; or the romance of the unexpected kick.

I suppose running from the rig floor where a couple of hundred MMcfd of sand-laden gas is escaping from the hole after blowing several joints of drillpipe past the crownblock could be considered the ultimate thrill.

It's the stuff that novels, movies, and morning-tour tales are made of.

But if drilling researchers have their way, making hole could lose much of what those located a safe distance from the rig floor have always called "glamour."

One article in this issue indicates, for example, that there are several methods for predicting formation pressures while drilling. Pressure-prediction techniques have reached a quite sophisticated level.

This development could make "drilling for kicks" a thing of the past.

Another story in this issue cites renewed interest by a number of firms in measuring drilling variables at the bit and transmitting the measurements to the surface.

Talk about taking the romance out of drilling! Imagine knowing what's happening at the bottom of the hole. It would no longer be necessary to consult with the stars from behind a mesquite bush or wave your hands over empirical charts and say magic words.

Yet another article in this annual drilling issue describes the latest rig for Arctic service. In addition to the most modern equipment available, its camp features a coffee room, recreation area, and movie theater.

A stint on a rig of this type, even in the Arctic, could hardly qualify as a really topnotch tale of hardship and deprivation.

Still, I have the feeling that many in the industry would not be too sorry to see some of these thrills disappear.

I wouldn't be at all surprised if many could learn to live with much more complete knowledge of what's going on down the hole, with adequate warning of possible pressure problems, and with almost-comfortable working conditions.

It would take some getting used to. But the industry has made adjustments before, and it may just be able to cope with the change.

Even science can be tolerable.

Paper mountain majesty

Leo R. Aalund **November 24, 1975**

Paper, most of it made up of federal reports, forms, and questionnaires collected by Congress and the agencies it has created, is becoming a national characteristic in the U.S.

It's estimated that the federal Government now has over 6,000 different forms in print. As a result, federal employees shuffle some 10 billion sheets of paper each year, enough to fill Houston's Astrodome 50 times.

These and other appalling statistics were presented at the recent National Petroleum Refiners Association computer conference in Houston by C. L. Zody, assistant controller of Exxon Co., U.S.A.

Though Exxon has always filed government reports, the burden grew immensely after Congress created a spate of new acronyms such as EEOC, FEA, EPA, and OSHA, which we won't bother to define here. They and existing agencies like DOT, SEC, and USGS were soon clamoring, often at the behest of their congressional masters, for more and more information.

Exxon saw the handwriting on the wall and early in 1974 started keeping track on the manpower and cost associated with its reporting load. As of now, the company finds it is filing 409 reports to 45 different federal agencies. This excludes all federal, state, and local-tax reports. Of the 409 nontax reports, 55, or 13%, were introduced in the past 18 months. These 409 reports required the equivalent of 112 employees at a cost in excess of $3.5 million/year. These are company costs not associated with the federal paper shufflers. On top of this, there are 17 new federal reports in the works.

State bureaucrats are jumping on the bandwagon too. Oregon wants annual 20-year projections by product. Finally, there have been eleven congressional questionnaires in the last 2 years.

Is there any possibility the country can tunnel out of this mountain of paper, which is costing consumers billions? There's one slim hope. That is the recently appointed Commission on Federal Paper Work. We hope its final report runs to less than 100 pages.

Changing of the guard

W. A. Bachman **January 5, 1976**

For the first time in more than 41 years, George Weber is not playing a direct role in the week-to-week editorial activities of The Oil and Gas Journal.

On Jan. 1, his dual role as editor and publisher of the Journal changed. He handed over the editor's job to Gene T. Kinney, who is moving to Tulsa after serving as Washington editor since 1964. George retains the post of publisher and will continue in the active management of all departments.

But this changing of the guard in the editor's office is something of a sentimental occasion. George Weber has been involved in writing or editing the Journal since 1934. In fact, he has been on the editorial staff longer than anyone.

George Weber also has been a contributor to the emergence of the Journal from an oil-field trade paper to a prestigious weekly magazine of international influence.

The reason George was hired in the first place was prophetic. He was the first graduate petroleum engineer on the Journal staff. Journal management back in 1934 recognized that application of engineering principles to operations in the oil and gas industry was the wave of the future. The Journal editor of that day asked the engineering dean at the University of Oklahoma to recommend a candidate who was a good student with a facility for writing. George Weber was recommended.

Jobs for engineers—or anyone else—were scarce in those depression days. So George decided to accept the magazine job offer, maybe until something better turned up. But he showed enterprise as an engineer-journalist, liked what he was doing, and stayed on.

George earned his reporting and writing spurs in the southwestern oil country. He did tours of field duty in Tulsa, San Antonio, Midland, Shreveport, and Dallas, and took time out during World War II to serve in the South Pacific as an artillery officer. George was named refining editor in 1947 with headquarters in New York City. During the next 12 years, he established the Journal's international reputation in processing and at the same time became one of the best known technical journalists in the industry. These accomplishments were rewarded in 1959 when he was named editor and transferred back to Tulsa. The role of publisher was added in 1973.

In the past few years, George has managed to take care of many administrative and writing duties, travel extensively, and in addition make scores of speeches before industry and public gatherings on all facets of the industry and its problems.

This little piece of recognition is certainly not intended to signal that George is being put on a shelf, because he's still a supercharged individual. But his colleagues in the Journal's editorial offices couldn't let his departure as a working editor pass unnoticed. They join me in tipping our hats to an "old pro" who has performed with distinction and proved himself a rare leader in a difficult field.

Bicentennial backlash

Max L. Batchelder **January 12, 1976**

Well, here it is at last. The bicentennial year.

It is hard to turn a TV knob or flip the pages of a newspaper or magazine without seeing some bewigged colonial character selling deodorant or used cars.

Some truly far-fetched tie-ins have been linked to the bicentennial theme (an essay contest on "The Role of the National Football League In American History!??"). A few serious efforts have been made, such as Shell's "200 Years Ago" television series, but mostly the bicentennial has been an excuse to peddle everything from almanacs to pizzas.

In an effort to correct this light-hearted approach to history, let me toss out a few historical notes oil advertising people might pick up on. Things that will add some dignity to the industry. Like the driller who shortly after Col. Drake's time, was moving his rig from one site to another a considerable distance away. An amateur spiritualist, he suddenly felt he was receiving a supernatural signal. A true believer, he immediately set up his rig, started drilling—and discovered Pleasantville field!

Or take the first pipeline company formed. When construction reached the railroad right-of-way, the railway refused to let them lay the line under their tracks (the railroads preferred to haul the oil themselves, you understand).

This produced a rather remarkable set of circumstances. The oil traveled by pipe to the edge of the railroad tracks, where it was pumped into tanks which were pulled by horse-and-wagon across the public highway to the other side of the tracks. There it was dumped back into the pipeline to continue its journey. The only extant photograph of this location shows what you might expect—a locomotive and a string of railroad cars stopped, blocking the intersection.

History is filled with these tension-packed vignettes little known to people outside the industry. We can tell how the Hamill Brothers, drilling at Spindletop, drove cattle through their surface mud pits to thicken the water with clay—the first additives to hold back unconsolidated surface sands.

We can tell the world how a driller in Texas went broke drilling a hole unsuccessfully and was about to pull up stakes when his wife pawned her last piece of jewelry and sent him back to drill a few more feet—and discover Burkburnett field.

Or the two geologists who refused to map a field because they feared ridicule for making it so large—the huge oil field under Oklahoma City which eventually produced millions of barrels of oil.

Or the first gravity surveys being made on frozen lakes in Europe. Or—but that should be enough for someone to put together some historically sound commercials that would lend a little stature to the industry—and you can always sell a little 40-wt motor oil at the end, of course.

Hollywood and the gas bill

Caroline Mayer **February 9, 1976**

"Go get 'em, John!" the congressional witness exhorted, thrusting his arm in the air for emphasis, as he was leaving a House committee hearing on natural-gas deregulation.

The battle cry was directed to the energy and power subcommittee chairman, John Dingell (D-Mich.), as encouragement in his fight against the campaign for deregulation. Normally, this outburst in Congress's staid committee rooms would have been as appropriate as a shouting match in a library.

But this was no ordinary hearing; nor was this a typical witness. No, this was an event extraordinaire—an event that brought the press and public out in droves. Even several congressmen who had been avoiding the subcommittee's past 2 weeks of hearings came to see the star, one of Hollywood's biggest box-office attractions, Blue Eyes himself—Paul Newman!

Billing himself as a "troubled fellow," Newman sought to speak as an ordinary citizen concerned with big oil's economic power and political clout. He came to Congress to glamorize the deregulation debate, which, he believes, had received too little attention in the national press.

His 25-min appearance, only 5 days before the House was scheduled to begin debate on the issue, was made on behalf of the newly formed, anti-industry Energy Action Committee, a group Newman and several friends helped get off the ground with an initial $600,000 donation.

"A scandal is brewing that may make Watergate look like kid stuff," Newman warned the hushed, star-struck committee. "And if that is true, the American people are likely to view those legislators who have mortgaged themselves to the oil companies as responsible as Mitchell, Haldeman, Ehrlichman, and Nixon were for Watergate."

But because Blue Eyes is no ordinary citizen, his testimony was heeded by the subcommittee members most attentively—more so than that given by the dozen other witnesses who had come before the subcommittee earlier.

For without star status, how could the words of a gas producer, a curtailed pipeline company, an industrial or agricultural user, or a Federal Power Commission official—all favoring some sort of gas deregulation—be important?

Then, too, Newman's words fell on friendly ears, for only those congressmen opposed to deregulation were there to hear him.

Newman's appearance floodlighted their cause, and his exhortation urged them onward. For if Hollywood was behind them, how could they be wrong?

A crude in every port

Leo R. Aalund **March 29, 1976**

Rocky Mountain Sweet and Alberta Mix are not two dancehall girls who fought for the favor of a cowboy called Texas Sour in a Deadwood, South Dakota, saloon one memorable night in 1886.

They are, of course, well known North American crude oils, which like many others are being overshadowed by imports, also bearing colorful names.

Foreign crudes are a major theme of this year's refining issue, which again has as its cornerstone the massive survey of U.S. and Canadian refineries. For both features, the Journal and this editor thank hundreds of industry respondents and contributors, here and abroad.

This worldwide flavor for a special issue that has traditionally focused on North American refining developments is not an editorial departure. Like their foreign counterparts, U.S. refiners must cope with a growing variety of world crudes.

Even the general public (many congressmen included) is now aware of the rising flood of imports. Week before last they hit 8.2 million b/d, the highest in history.

This development seemed to have a bigger impact on this editor's circle of non-oil industry friends than any oil news since the embargo. Though these people, like those in industry and government, are mystified by entitlements, they clearly understand headlines that say:

"U.S. gets half its oil overseas"

Even political candidates can't fog up that one.

So, until the U.S. starts rationing or gets out and develops its own energy sources, the amount of foreign crude going into U.S. refineries will just grow and grow. A guide to world crudes is appropriate to the times.

Making this compilation of export crudes required some grinding drudgery, but it was offset by the fun of looking at the world in light of oil geography—particularly for an editor, who as a teenager, dearly wanted to ship out on an Esso tanker bound for the Persian Gulf. But career guidance at the docks was quite straightforward in those days:

"You go home, go back to school, and don't come back down here again."

Well, there's more than one way to get to Ras Tanura, Brass River, Puerto Miranda, and even Valdez in Alaska. They and other oil ports, as this issue shows, open the door to the world for export crudes.

A sad tale: Project Retrospect

W. A. Bachman **May 17, 1976**

Whatever happened to Project Independence?

More than 2 years have now passed since the Government launched its bold effort under that title to make the U.S. self-sufficient in energy and free Americans from the threat of future OPEC oil embargoes. But somewhere we've wandered off course.

How we went astray is interesting, even if sad. The Journal is indebted to reader Thomas L. Wright of San Anselmo, Calif., for a recent letter in which he outlined the various transmutations which Project Independence has undergone. Here's his catalog of changes:

Project Indifference—predecessor to Project Independence. Warnings of impending domestic shortages were ignored or countered with the threat to allow increased imports of "cheap" Mideast oil if U.S. producers raised their prices.

Project Interference—the gasoline allocation system which led to—

Project Inconvenience—blocks-long queues of cars lined up at service stations.

Project Inquisition—Congressional grilling of oil-company managers by would-be presidential candidates with the 10 o'clock news in mind.

Project Insufficiency—preferred by environmentalists who would be happy with it until they ran out of firewood and bicycle tires.

Project Insignificant—unwarranted hopes for remote, unproved and/or far-fetched alternate energy sources.

Project Incoherence—instead of easing price controls on natural gas, politicians are providing an alternate supply. It has a high heat content, but no illuminating qualities.

Project Inefficiency—Fogco, the proposed federal oil and gas company. Studies of government-run oil companies in other countries, made by the FEA, show that private companies produce an average of 44.2 b/d of oil per employee, compared to 6.6 b/d per employee for the government companies. In refining, the output of private firms averaged 36.4 b/d per employee compared to only 11.4 b/d for the government firms. The consumer pays at the gas pump for this inefficiency.

Project Incapacitate—the current clamor to disintegrate the larger oil companies into segments which would be prohibited from coordinating the continuous flow of petroleum from oil wells through refineries and into our gas tanks.

Project Inconsistent—claims that the oil companies caused the U.S. to rely too heavily on oil and gas resources and now should be barred from diversifying into developing alternate sources of energy.

Project Inequity—an OPEC or other foreign producer can sell his oil to Europe at $12/bbl, but an American producer must sell his (old oil) domestically at $5.25/bbl.

Project Insolvent—we are now sending the OPEC countries more than $2 billion/month to pay for oil imports.

Project Indecision and Project Ineffectual—current administration versions of Project Independence.

And Mr. Wright might have added "Project Dependence—the final transmutation."

If his success be not now, perhaps it will come.

The readiness of the Renaissance man

Caroline Mayer **May 31, 1976**

Regulating the price of a valuable and dwindling commodity is like "paying Picasso on a cost-plus basis for his paintings," Rep. Robert Krueger (D-Tex.) told his colleagues on the energy and power subcommittee last year when they were debating oil-price controls.

Beginning with that analogy, the freshman congressman has emerged as one of the chief spokesmen for decontrol—one who, almost single handed, has taken on the industry's most precious cause—natural-gas deregulation.

In so doing, the 40-year-old Oxford scholar and former university professor has distinguished himself not only from the liberal-minded freshmen who entered Congress with him but also from most of his senior colleagues.

For instead of spouting political platitudes, the tall, composed Renaissance man uses Shakespeare, Greek mythology, German, and poetic metaphors to win converts to his cause.

Success hasn't been easy, though. He's been narrowly defeated on the issues for which he has fought the hardest. And to make matters worse, his erudite thoughts are often lost by the baffled congressional record keepers who fail to understand him.

For example, on one occasion in the energy and power subcommittee, Krueger expressed the hope that Chairman John Dingell (D-Mich.) would be a "Rhadamanthine judge"—one who, in Greek mythology, ruled in an extremely stern, but fair manner.

Not only, some think, did Dingell prove otherwise, but this profound thought so confused the stenographer that the expression never found its way into the hearing record.

Nonetheless, this articulate champion of a not too popular cause perseveres—trying to win votes through his refreshing, poetic charm.

To illustrate:

On charges that producers have shut in their gas to win higher prices in the future: "This search for shut-in gas reminds me of a search for the Holy Grail."

On the reason why deregulation advocates in the Senate are supporting the compromise gas bill: "Impatience can drive good men to desperate acts."

On the haste with which the Senate commerce committee approved the compromise gas bill: "They didn't even grease the skids; they just took a straight vertical plunge."

On whether gas deregulation would occur this year: "If it be now, tis not to come; if it be not to come, it will be now; if it be not now, yet it will come. The readiness is all." (Hamlet, Act V, Scene II).

Armed with his dry wit and lyrical manner, Krueger is ready—ready to prove that oil and gas should be treated as valuably as a Picasso.

The Paul Getty we knew

George Weber **June 14, 1976**

It may seem strange to some people that J. Paul Getty will not be remembered simply as the world's richest man. But it's true. To us, he was known mostly for his other accomplishments and for the varied viewpoints he shared with us over the years.

First, we regarded him a regular, long-standing, and very careful Journal reader. And he made certain that we did. He seemed to enjoy reminding some of us he had been reading the book longer than we individually had been writing it. His thorough awareness of our coverage of industry news, interpretation, and technology proved it.

Paul Getty never quite left Tulsa and the Oklahoma oil patch where he got his start and with which he always warmly identified. Some of that rubbed off on us, and we would be amused and pleased at the strong proprietary feeling he held for the Yellow Book, without ever owning a nickel's equity in it. Not always were we amused, for he could deliver a real dressing down for our sins of omission. The situations he wanted us to "straighten out" were legion and formidable.

Next we thought of him and will remember him as a wildcatter in the broadest sense that name implies. His was the breed that found and developed the almost boundless energy a world consumed without counting the cost, and came to take for granted. To the end of his many memorable days, Paul Getty remained the explorationist who resisted the urge to go back to sunny California, and stayed for one more gray chilly winter in Britain—just to keep his hand in on the North Sea. Too bad, but that was the way he was.

We may be the smallest of minorities in regarding this man as a conservationist with a long vision of the distant future. Way back in 1919, he began to apply heat to his sizable reserves of heavy oil in the San Joaquin Basin. The early tests were crude and not too effective, but he kept at them over the years, perfecting techniques that were to come to attention dozens of years later. During that period he was spending several dollars to recover each barrel of sour, heavy oil that then sold for $1.25. Today, this foresight is paying off handsomely for Getty Oil Co. But the advanced thermal-recovery know-how will profit the oil industry and energy consumers as well. Billions of barrels of oil will be recovered using tools and practices developed over the past 57 years.

We know this firsthand, for we were proud to publish for the first time reports of Getty's thermal operations, with his knowledge and full approval.

His passing came just 3 weeks after he was honored at the International Petroleum Exposition in Tulsa as "Pioneer of Pioneers," a fitting tribute from old friends which touched him deeply.

Reader, wildcatter, conservationist. We wish the general press had had our vantage point from within the oil business and could delineate his life with facts more meaningful than the wealth he could not count, and the many trivialities that still surround his image.

But they didn't know the Paul Getty we knew. It was their loss.

The 'Minas story'

Larry Auldridge **June 21, 1976**

Mixtures of people, purposes, and circumstances have contributed to many interesting highlights (and footnotes) to the chronicles of world petroleum.

One such example involves two geologists—an American and a Japanese—at a time when they were young men worlds and armies apart. But both had roles in the discovery of the huge Minas oil field on Indonesia's island of Sumatra.

It's a fact, but not widely known, that Minas field was actually discovered by Japanese drillers in September 1944, while Indonesia was under occupation in World War II. The Japanese used a rig and equipment left behind by Americans evacuating the area ahead of invading troops.

The site geologist was Turo Oki, a geology graduate of Kyushu University, who then was in the Japanese army. He still was there in 1945 awaiting repatriation.

Richard T. Hopper, Oki's counterpart, was a geologist for what is now P. T. Caltex Indonesia Ltd. before the war and returned to Indonesia in 1945 after discharge from the army.

As Hopper relates it in a recent issue of Caltex's company publication, he learned of the 1 Minas oiler, but was barred for military reasons from traveling to the remote jungle area.

The Japanese had produced "several thousand" barrels of Minas crude for a nearby small distillation unit.

In response to a request from Hopper, the Japanese geologist provided by courier two 5-gal cans of crude, core samples, and a log that were shipped to the U.S. But Oki returned to Japan without meeting Hopper.

In the ensuing years, Caltex (jointly held by Texaco Inc. and Standard Oil Co. of California) developed Minas field. It has produced nearly 1.9 billion bbl of oil and currently has an output of 400,000 b/d. Japan is one of the largest importers of Minas oil.

However, the "Minas story" didn't come to full cycle until recently. Hopper, now vice-president of American Overseas Petroleum Ltd. (Amoseas), also a Texaco-Socal affiliate, just last year met Turo Oki in Tokyo.

Turo Oki now is managing director of Japex Indonesia Ltd., an affiliate of Japan Petroleum Exploration Co. And oil again was the business at hand for the two oilmen. Japex entered a joint exploration venture off East Kalimantan, Indonesia, with units of Amoseas' parent companies.

Once again after three decades, the two geologists share the same interests—not the least of which is that quite unofficial if not unorthodox short-term Minas field "joint venture."

On big and little money

W. D. Moore III **August 16, 1976**

The United Arab Emirates is a small coastal enclave on the vast Arabian desert. The climate is harsh, but the people are extraordinarily friendly. Arab hospitality shown by both nationals and expatriates is superb.

I thoroughly enjoy a trip to the U.A.E. Whether visiting friends of many nationalities or loading up on bargains from the souks of Dubai or Abu

Dhabi, it's always a lot of fun.

The money in the U.A.E. is the dirham—a solid currency with an exchange rate about $0.25 U.S. to one. Paper money starts where coins leave off, at one dirham.

A most common coin is an undistinguished-looking copper-nickel piece about the size of a U.S. quarter. It's hard, durable, and not very pretty. It's the half-dirham, worth maybe 12¢ U.S.

The obverse (heads) of a half-dirham coin is not very memorable. The country is named in two languages, and the denomination is shown in Arabic.

But the flip-side of the coin tells a different story. If it's not memorable to you it should be. The back is adorned with three drilling rigs.

I don't know how many countries put rigs on their money. The U.S. doesn't. The U.A.E. does. There are three rigs on the half-dirham, and at least a derrick on every denomination of dirham bill.

The U.A.E. is not alone in this. If you flew to Bahrain, you could swap 2 1/2 dirhams for a note bearing both a land drilling rig and a refinery. But the U.A.E. probably ranks first in engraving the source on their money.

This little coin typifies attitudes of Middle East businessmen toward oil. They like oil. They favor producing it. They want to make money from it for themselves and their country.

Imagine what this attitude would do for U.S. production. Imagine the chance of Congress honoring oil on a coin. Imagine where your company should be doing business.

Oil revenues will bring the U.A.E. billions of the little oilwell dirhams to spend. They are buying houses, schools, roads, rigs, rags, rugs, and pipe rubbers, not to mention rope, soap, and dope. It's all imported.

As the reader can see from the special report beginning on p. 75, the U.A.E. and other Middle East oil producers like new technology and its products. They are converting dirhams, dinars, rials (or riyals), and plain old dollars into real wealth that means a better life for their people.

They are creating unprecedented opportunities for those willing to do business in a different place and manner.

Low-lead foot

Leo R. Aalund **August 30, 1976**

For almost 2 years you have been slightly frustrated, either obeying the law and driving at 55 mph or going just a little faster and feeling like a cheater.

But the temptation was there. It's the late afternoon of a beautiful summer day and you're driving a new car with the radio music turned up high.

There's nobody ahead or behind and a magnificent stretch of concrete slopes gently downward for miles through a forest of stately pines. Why not, just once, go as fast as you want?

Some downward pressure on the pedal and the needle glides down to the 90 mph region on the speedometer. That's exhilarating, so you savor it awhile. Then you put on a little more foot, and you're on your way to 100.

But then it dawns on you that you're not hearing the music and are gripping the steering wheel with both hands. You're out of your element, or perhaps chicken, and drop back to a sensible speed.

But that fun was all legal because it took place on a West German autobahn on the outskirts of Frankfurt headed for Ingelheim on the Rhein. With few exceptions, there are no speed limits on the superhighways which sweep to the points of the compass all over West Germany.

Autos flashing by at 110 mph are not uncommon, and newcomers quickly learn to never, never, spend more than a few seconds in the passing lane—if they can find anyone to pass.

Some Germans are dead set against this unbridled velocity, which is a political issue. But others insist that a speed limit would be an abridgment of personal freedom. But the root reason for not having a speed limit, anti-speeders say, is that West Germany builds and exports some of the finest and fastest motorcars in the world. How would it look if the exporting country made a product that couldn't be fully utilized in the land where it's made?

As a result, German refiners are getting it in the neck. The West German Government apparently feels lead in gasoline is a greater danger than speed. As of Jan. 1 this year, German refiners had to cut the lead level to a very low 0.15 g/liter or 0.6 g/U.S. gal. But on top of this they have been ordered to continue making 60% premium and 40% regular—as the auto makers of Stuttgart and Munich keep turning out high-compression engines.

For the moment, German refiners have a tougher problem than their U.S. counterparts. The research-octane number of German premium is 98, with only 2 or 3 numbers coming from the permissible lead. This low level of lead makes it likely that a motor-octane number of 88 will be required, German refiners tell this editor.

Those are bad numbers in any refinery.

The Flug anxiety syndrome

Caroline Mayer **September 27, 1976**

When the antioil lobbying group Energy Action entered the political arena nearly a year ago, the oil and gas industry quickly came down with a severe nervous disorder. The mere mention of the group's name made some industry representatives more than a little nervous.

With good reason. Energy Action soon became one of the most vocal, articulate, and energetic antioil organizations around. Although new, it was no novice in the art of getting good publicity for its causes—divestiture and continued natural gas regulation.

In fact, the organization had it easy. All it had to do was promise to produce one of its cofounders, Hollywood box-office attraction Paul Newman, for a press conference. Reporters would turn out in droves, leaving little room for the star himself.

But that was several months ago. No longer do industry representatives steam with frustration upon hearing the words "Energy Action." Now they just chuckle. For in the past few months, the group has succeeded in alienating both the press and numerous congressmen who would normally be its allies.

The press was the first to get disenchanted after Director James Flug explained to a television audience last July why so many U.S. newspapers are editorially opposed to divestiture: "Ask them (the newspapers) how much petroleum-company advertising they have," he said, adding incorrectly, "They're the biggest customer."

The anger of an ardent reregulationist, Sen. Adlai Stevenson (D-Ill.) was soon to follow, after Flug accused him and his colleagues of succumbing to a "snow job" by the natural-gas industry by believing that higher prices were needed to bring on more gas supplies.

Appearing before Stevenson's subcommittee on oil and natural-gas production, Flug implied that the industry had intentionally drilled dry holes off Florida's coast after it had spent hundreds of millions for a federal lease. He attacked his favorite target, Mobil Oil Corp., charging it with "chutzpah" because it advertised warnings of an impending gas shortage while it "purposefully keeps gas off the market." And he accused the industry of buying congressional support of deregulation.

Incredulous and incensed at Flug's testimony, Stevenson dismissed him by pointing to the stacks of commerce-committee-hearing records on natural-gas pricing and by telling him to go do his homework.

"We were lucky last winter. It wasn't just (a result) of faulty administration calculations" that we didn't have a severe gas shortage, Stevenson told Flug.

Upon hearing this exchange, the industry's anxiety vanished. For the industry realized that the irritant causing it had neutralized itself.

The Houston converter caper

Leo R. Aalund **March 28, 1977**

If you like "man bites dog" stories, you will be pleased to hear that the cops of the city of Houston have just been ticketed.

They face a $46,750 fine.

Like U.S. refiners, the Houston police don't care for catalytic converters. And being men of action, they sawed the offending elements off their new patrol cars in their own shop and put on straight pipes.

When the Houston Post blew the whistle on them, their excuse sounded sadly unimaginative. They not only didn't know they were violating the law but believed the Environmental Protection Agency had told them it was all right to take off the converters.

The first impulse is to snicker—until you start looking into the ludicrous status of catalytic converters.

Texas is a good example of what is probably going on in all but a few of the "United" States.

The Texas Department of Public Safety (TDPS) does not require that the thousands of licensed commercial facilities performing annual auto inspections check emissions. The inspection fee set by the legislature is too low. However, a TDPS official in Austin tells this editor that the private inspectors are supposed to nail anybody that has disabled or removed a pollution-control device.

"Well sir, is it possible to determine if a converter is working without an emission test?"

Not very well, he admits. But he won't venture an opinion on how many of the private inspectors even know what autos are supposed to be equipped with converters in the first place, what with light-duty trucks, foreign cars, etc.

And there is no federal law, the EPA tells this editor, that forbids a car owner—not dealer—from removing a converter. That's where the Houston cops apparently got the wrong idea when they asked.

But the Texas Air Control Board (TACB) has promised the EPA to enforce a state antitampering regulation. But the TACB doesn't have any converter inspectors and is obviously in the dark if it is depending on the TDPS. If the TACB doesn't enforce this reg, the EPA can step in and do it.

At the moment, the TACB, which had this case handed to it on the front page, doesn't plan any punitive action against the Houston police, just teach them the straight and narrow. If you get stopped by a cop in Houston, you may want to cite this precedent.

But the police made another, bigger mistake. They put leaded gasoline from their own pumps into their formerly converter-equipped cars. That is a federal offense and the reason the feds want almost 50,000 big ones. The city of Houston will appeal this fine, which could have been much steeper.

Meanwhile, the nation's refiners build, sweat, worry, and spend money to make unleaded gasoline.

The lesson of Leduc

Jim West **May 23, 1977**

For the past 6 years or so, it's been my pleasure to help cover the Canadian oil industry for the Journal.

It's a great assignment. The Canadian scene is never placid. In fact, sometimes it's downright volatile when the industry is caught between provincial and federal governments squabbling over which has the power to tax and regulate oil and gas.

Through it all, however, the Canadian industry rates among the most cooperative in the world when it comes to providing the Journal with information it requests for a story. I've never met most of them, but I feel a strong kinship with Canadian oilmen.

So I'm particularly pleased to help note the birthday of the giant field that launched the Canadian industry as we know it today. And in that birthday celebration lies a good lesson for energy officials of the Canadian and U.S. governments.

It was 30 years ago that Imperial Oil Ltd. discovered Leduc field.

The dramatic strike on the Alberta prairie quickly turned the Canadian industry around. Things had looked grim indeed before then.

Imperial Pres. J. A. Armstrong recalls that in 1947 Canada's oil industry, in spite of being 90 years old, was still in diapers.

Domestic oil production amounted to less than 20,000 b/d, a mere 8% of demand. The nation relied on imports from the U.S. for the remaining 92% of its oil supply. Experts were warning that world demand threatened to outstrip supply.

Despite a long and costly exploration program, Imperial had made no strikes of any consequence since discovery of Norman Wells field in 1920.

"In western Canada," Armstrong says, "Imperial had established the kind of record we would rather have done without. We had drilled 133 consecutive dry holes.

"We were ready to give up exploration. In fact, we were working on plans to synthesize gasoline from natural gas.

"But explorers have a penchant for drilling one more well."

That one more well, as it turned out, was 1 Leduc, a 600-million-bbl discovery that set Canada on the road to energy self-sufficiency.

The U.S. had a whole batch of giant fields under its belt at the time of the

Leduc strike. But Leduc is especially important because of its significance to an entire country. What's more, it shows what explorationists can do, given adequate incentives in Canada, the U.S., or anywhere else.

The need for exploration incentives is a lesson that has to be learned by every new bunch of lawmakers, committee counsels, energy advisers, systems coordinators, and assistant deputy directors that invades Ottawa and Washington.

And if these and other government officials—clear to the top—don't know how to solve the energy-supply problem, they at least should stay out of the way of oilmen who do.

Dutch butter, sideways snow, and Catch-22

W. D. Moore III **June 6, 1977**

A bunch of good things are happening in the North Sea. It's a fine oil patch. A bunch of mildly curious things are happening, too. Some are closely oil-related, some are not.

In Holland, I decided Dutch butter is the best there is. Texas butter isn't bad, but Dutch butter's better. And due to some quirk in the Common Market there is certainly plenty of it. Governments put the price way high, thus promoting some really large-scale production. Even surpluses. What do dairy farmers know that we don't?

In Stavanger, I received an unexpected appointment with the 100-year storm. Reports from offshore said 20-m waves and 100-knot winds were slamming into the new Frigg platforms. On land, it snowed horizontally. All helicopters were grounded, and that suited me fine.

Actually, this was not the centennial storm of current predictions. Operator's initial estimates were blown away when the centennial occurred almost once a year. This particular weather may have been only a 1-year storm, but it was bad enough, and entirely too frequent. The very best times would not have been washdays, even in Midland.

From London, I planned to fly home Apr. 8 in time for Easter. My restricted excursion ticket was good Friday, Good Friday. It was not good Saturday or Sunday. The ticket could be used again Monday, but expired Tuesday.

Now British Airways suffered an "unofficial" strike the week before all this. Something about the tea-trolleys not being close enough to the work area. Planes would not fly Friday, and maybe not Monday. Some might get out over the weekend but this wouldn't help me without a new ticket. They weren't selling new tickets. Not until the planes were flying.

"No problem," says I, "I'll just go see Pan-Am." But since this strike was "unofficial," BA could not endorse my ticket over to another airline. (They would have in the event of a real strike.)

They further pointed out that my excursion discount would be lost if I didn't fly the BA flight that wasn't flying. Unless, of course, the strike became "official."

Under this Catch-22, I faced the choice of (1) junking a perfectly good, paid-for ticket, or (2) hoping a real strike would cancel all BA flights.

Confusing? You bet. I didn't understand it then and I don't now. But BA said hang in there, so I did and it worked.

I have a feeling someone gave North Sea operators similar advice some years back, Facing technical, economic, and political challenges, North Sea oilmen hung in there.

Progress detailed in this issue's special report is evidence that this attitude paid off.

$10.6 billion? They can't be serious!

John L. Kennedy **July 4, 1977**

I'm not sure, but I feel I'm going to be had. And if I am, so are you.

Let me explain.

Pick a number like $3 billion. About that much was spent on oil and gas in Saudi Arabia last year where one of the most grandiose development schemes is under way.

Or take an arbitrary figure like $3.5 billion. That's what the Journal estimates the entire Canadian petroleum industry will spend in 1977.

And when you need to make a point, there's nothing like the Alyeska pipeline, "the largest private construction project ever." At last tally, it will run in the neighborhood of $7.5 billion.

Of course, the petroleum industry is not all work and no pay. Net profits of the five largest U.S. oil companies in 1976—from all operations all over the world—was about $6.15 billion.

None of this, however, is the basis for my current suspicion. What does worry me is this matter of $10.6 billion. That's the first year's proposed budget for the about-to-be U.S. Department of Energy (DOE).

"$10.6 billion? They can't be serious!", you say. Oh, but they can.

"Yes, but..."

Stop stammering and listen. I know we only spent $6.6 billion drilling and equipping all the wells drilled in the U.S. in 1975. And I know it only took a meager $2 billion to drill and equip all the gas wells in this country that year.

I also know the U.S. industry, during 1977, will shell out just $1.55 billion for refining projects; only $2.02 billion for all its petrochemical-related projects; and a paltry $1.46 billion for crude and products pipeline construction.

"Are you sure it's $10.6 billion, not $10.6 million?" you ask hopefully.

Yes I'm sure, you dreamer.

And get this: All companies will spend only $10.2 billion on all drilling and exploration in the U.S. during the whole of 1977.

I must admit, however, that I found a number larger than the DOE's proposed first-year budget. It's the total estimated capital and exploration expenditure in the U.S. by the entire industry this year. That number is $28.45 billion.

The ecstasy of that moment of discovery was forgotten, however, when it occurred to me that the DOE budget is equal to 37% of the industry's total outlay in the U.S. And that's for the first year. If you believe it won't increase the next year, you'd better took under your pillow in the morning. Someone may have left you a quarter.

A quick-and-dirty conversion indicates U.S. production of oil and gas was about 17.7 million bbl of oil equivalent in 1976. And $10.6 billion/year is roughly $29 million/day (you government economists check me on this).

The capital outlays ($28.45 billion) needed to find, drill for, produce, transport, and process oil and gas will cost the consumer about $4.50/bbl of oil equivalent this year.

For an additional $1.65/bbl of oil equivalent, the consumer is going to get the DOE.

It will do none of these things.

Happy anniversary to us!!!

Carl J. Lawrence **August 15, 1977**

Subscribers to The Oil and Gas Journal are in for a surprise.

The Journal is celebrating its 75th birthday by giving you a gift—an extra issue of the Journal entitled "PETROLEUM/2000." We'll admit our gift is not like getting a new set of golf clubs, but we think you'll find it special and useful. In fact, we think you'll recognize "PETROLEUM/2000" as an outstanding reference piece that should be part of your permanent library.

Like all other issues of the Journal, "PETROLEUM/2000" is a joint venture between Journal editors and our friends in the oil patch. Admittedly our Diamond Anniversary special is vastly different from the first Journal published on May 24, 1902, in Beaumont, Tex., but our basic reporting philosophy of staying close to the industry is just as evident today as it was 75 years ago.

We could recommend some articles we feel are "must" reading in "PETROLEUM/2000," but given the lineup of authors we wouldn't know where to stop. Rather than single out any articles, let's just say there is plenty there for everyone. How much of the issue you read will depend on your interests.

"PETROLEUM/2000" covers the same wide range of subjects you find in your regular issue of the Journal—technology, news events, economics, statistics, trends, and plenty of projections. There's some history, but most of the issue is devoted to what will happen in the oil patch between now and the year 2000.

There also is a variety of authors. Most of the Journal staff wrote for the issue. Those of us who don't have a byline article in "PETROLEUM/2000" did other things that helped get the 542-page issue to press. And our friends in the industry contributed some excellent articles. You'll recognize some of the contributing editors as C.E.O.'s, some as association leaders, some as pioneers, some as thought leaders in research and technology, and some as guys who long have been recognized as the best informed in their field.

We're in the final stages of printing this special issue. It should go into the mails any day now. The exact delivery date is hard to pin down since this is a 2 1/2-lb package that requires special packaging and handling. But it won't be long until you have your copy.

The thing you may like most about "PETROLEUM/2000" is the common thread that runs from cover to cover—that oil and gas will be major factors in the energy picture in the year 2000 and beyond.

Here at the Journal we feel so strongly about the future of the industry that we're looking forward to the year 2000 rolling around. We know we'll be facing a decision then on what to say in our 100th anniversary issue. I'll predict that if we have the kind of staff and friends then that we have today, that some editor will be at his typewriter 25 years from now telling Journal readers how great the next big birthday special issue will be.

The first barrel

Gary E. Congram **August 22, 1977**

On July 28 at 11:02 a.m. (Alaskan Daylight Time) a significant event occurred in Valdez. After years of literally blood, sweat, and tears, the first barrel of crude arrived at that city from Prudhoe Bay.

Its arrival climaxed the dreams of thousands who had braved Alaska's cold winters and rugged terrain to build the most expensive pipeline in U.S. history. I was fortunate enough to be on hand for the delivery of that first barrel to Valdez.

It is difficult to comprehend how much effort has gone into producing and delivering that first barrel of crude without having seen Alaska and its people first-hand. Alaska is majestic and dynamic on the surface. But underneath is a sleeping giant which can command nature's mightiest furies.

Many of those furies were encountered by the builders of Alyeska pipeline. And the men and women of Alaska who built the line as well as other residents of this state are perhaps the most interesting part of this frontier. They are not losers who are making executive-type salaries, but rather winners who are seeking to free us from the shackles of our present energy shortages.

Although some writers have slandered the total cost of the pipeline ($7.7 billion), and lately some have proclaimed a needless cost of $1.5 billion due to poor management and waste, I feel that every penny was well spent for this project.

To sum it all up, I am reminded of the opening lines of Robert Service's poem "The Cremation of Sam McGee":

"There are strange things done
In the land of the midnight sun
By the men who toil for gold
And the Arctic gales have their secret tales
That would make your blood run cold."

Strange things indeed—such as building an 800-mile pipeline that moves black gold.

The Washington language barrier

Ms. Dale Tate **September 5, 1977**

If the public has a hard time understanding the energy crisis it's no surprise to me.

I've been listening and watching our national leaders the past several years debate the "big" energy issues and all the minutiae that goes along with such debating. There are many times when I am slightly confused.

It's a well-known fact that Washington, D.C., is the bureaucratic jingo, jargon, acronym capital of the western world. And when I started covering the oil and gas scene, I knew I would have to learn some special terminology.

So, I set about my task and learned the difference between crude and refined oil (coming from Boston I figured that one out fast) between an independent and a major, and what the initials OCS, FEA, FPC, MER, MPR,

MMT, EIS, EPA, NPR, MFC, TCF, SPR, NGL, SNG, and OGJ all stood for.

But that was just the beginning. It's been a never-ending educational process and every time the administration—be it Nixon, Ford, or Carter—sends a new energy proposal to Capitol Hill, I must add a new glossary of terms to my already ponderous vocabulary.

I know I must, but do I really want to know the many nuances which distinguish "old" oil, which is lower tier or first tier from "new" oil which is upper or second tier, and still to come "new new" oil, though not yet finally defined, will most assuredly fall into your third-tier category.

And now the National Energy Act, which President Carter sent to Congress late in April, has loaded me down with a whole new raft of catchwords and brought back some "oldies but goodies."

Instead of "decontrolling" oil prices as the industry was hoping, the Carter administration proposed a "crude-oil equalization tax," more affectionately known as COET.

But as soon as the tax was mentioned, congressmen representing oil-producing states floated that good ole "plowback" investment-tax credit which would certainly be tied to a "windfall-profits tax."

These taxes no doubt would have a substantial impact on the economy so one had to use the "arithmetic mean" to determine whether or not the "COET" would have an "updip" or a "downside" effect on the "GNP inflator." One certainly hoped not, but when you're dealing with the "BTU equivalency price" it's hard to tell.

Once the discussion turned from production to conservation the language became just a bit seamy—there was lots of talk about "cogeneration" and even "trigeneration." The terms "retrofitter," "economizer," "turbulator," "heat exchanger," "heat pipe," "heat pump," and "heat wheel," not to mention the "harmonic mean" (which is the inverse or the reciprocal of the average gal/mile) were bandied about freely.

Deregulation of "new" natural gas, whether it came from "geopressurized brine" or is "wet" or "dry," was heatedly debated and will continue to be so ... along with the "entitlements" program, the "small refiner bias," and "indicated," "inferred," and "implied" reserves.

Catch my drift?

The more the merrier!

Randall Scott Sumpter **October 31, 1977**

Energy—particularly oil and gas—is becoming a hard story to top in the daily press.

It wasn't long ago that only oil patch and big-city newspapers covered the industry's news. Even then, most items went on inside pages or the business section. But the energy crisis changed all that.

Energy is now front-page, above-the-fold, banner-headline NEWS. And that's had an impact on the job we try to do at the Oil & Gas Journal.

Less than 4 years ago, a Journal editor on an assignment would find himself all alone at the press table or interview. Now there's a crowd. A recent press conference I covered in Tulsa drew a veritable pack of news hounds: three television camera crews, six radio and newspaper reporters, and the local United Press International correspondent.

There are other signs that the great energy-news rush has become a stampede. Journalism educators put more emphasis on business reporting, professional magazines give recipes for successful energy exposes, and television specials and in-depth newspaper series on the who, what, where, when, why, and how of energy abound.

Even the Associated Press's revised stylebook—the final word on style, diction, and grammar for most journalists—is afflicted. It gives a complete listing of the OPEC member countries, the correct way to handle Arabic names (Ahmed Zaki Yamani is cited), and how to abbreviate some of the big multinationals' names (Aramco, ARCO, BP, Socal, and a few others are listed).

There's even an entry on how to convert tons of crude into gallons. Of course, the example of how to work the arithmetic is based on a "how many gallons in an oil spill" situation.

The results of this great rush to cover energy news are often amusing—and frequently inaccurate. A few examples: A national commentator refers to the "great Permian field." Cost of the trans-Alaska pipeline is variously reported as $9 billion, $9.6 billion, $97 billion, and $7.6 billion. And "flags of convenience" tankers continue to receive the blame for most oil spills when a recent study by the Tanker Advisory Center shows that it just isn't so.

So what does all this mean for the Journal? If anything, the competition has made many of our readers more eager to get the next issue to see what actually did happen. More than that, it has influenced us to dig deeper, do more interpreting for our faithful, and more reporting on what comes next. We've been called the bible of the industry, and we like it. We plan to keep it that way.

An offshore milestone

Bob Tippee **December 12, 1977**

When the Ship Shoal platform went to work 30 years ago in the Gulf of Mexico, 45 miles south-southwest of Morgan City, La., its operators were trying to conquer a frontier they weren't even sure was there.

Kerr-McGee's Ship Shoal Block 32 No. 1 was more commonly called in its day "the gamble in the gulf." The platform cost $230,000 to build, and when it spudded its first well, no one knew whether the ocean bottom contained enough hydrocarbons to even begin to pay out the costs.

On Nov. 14, 1947, the gamble in the gulf won the pot. At about midnight, it flowed sweet oil at a rate of 40 bbl/hr through a 1/2-in. choke.

The pioneer became the world's first commercial offshore well out of sight of land. For the industry it spawned, it's been a run for the roses ever since.

Today there are an estimated 1,250 multiwell platforms active around the world, the biggest of which cost more than $300 million to build.

They are no longer simple derricks fitted with sealegs and serviced from barges tied alongside. Many of them are self-sufficient habitats that can drill to seemingly limitless depths, conquering a frontier no less formidable than outer space, in the face of regulatory obstacles undreamed of 30 years ago.

And they're doing it for a reason that's at least as important as finding out what's on the moon's backside: About one-fifth of the world's hydrocarbon reserves lie offshore. And more than 80% of the offshore reserves are believed still in place, compared with about 50% of onshore reserves.

The offshore industry's technology explosion is obviously bigger than the imagination of a new Journal staff member who recently learned the difference between jack ups and semisubmersibles.

But even a novice a couple of years shy of the Kerr-McGee milestone can understand one thing: The offshore industry is one of the principal reasons the lights will stay lit in the 1980's and beyond—if regulators allow the industry to do its job.

Why not another Richter scale?

Patrick Crow **April 17, 1978**

One of the frustrations of being an editor is noticing an underlying trend in a series of news stories but not getting the chance to comment on it in print.

But this column is my own this week, so here goes:

Although the recent Amoco Cadiz mishap off France was the worst tanker spill in history, its treatment in the world press (although not the French press) was relatively light. The Torrey Canyon spill across the English Channel was only half as large but drew bigger headlines, as did last year's Ekofisk field blowout.

"The Amoco Cadiz spill may have been the biggest in history but it wasn't the worst," an Amoco spokesman explained.

That depends on which side of the oil spill you're standing, but he had a good point: the crude was light, constant storms dissipated it nicely, and there was a limit to the potential spill — unlike the Ekofisk cliffhanger.

I'm surprised there isn't a rating method, sort of a Richter scale, to measure the magnitude of oil spills. It could include such factors as volume, crude type, climatic and sea conditions, proximity to shore, and the like. Such a composite grading system might counterbalance the hysteria of the general press when it sees rainbows of oil on water.

That and other such self-evaluation would build the industry's credibility in regions where it needs credibility the most — off the U.S. East Coast, for example.

Perhaps the discovery of any commercial production off the Middle Atlantic and New England states would prove more important than the actual reserves found, unless they're of the Prudhoe Bay magnitude.

We all know how Texas and Louisiana feel about the oil industry. But will offshore production have a positive effect (from industry's point of view) on the psyche and politics of the northeastern states?

Maybe if that population is injected with a sizable oil-industry workforce, it would become immunized against "big-oil" prejudices, and some of its vociferous anti-industry congressmen would flipflop.

The Rhode Island example lends hope to this theory. That state, in its quest to become the primary service and supply base for the Offshore East Coast oil industry, has taken a more moderate view on Continental Shelf drilling.

From the public relations standpoint, where there's oil (production), there's hope.

Technical humor

L.R. Aalund **May 22, 1978**

The refining and petrochemical industries are populated with a large number of diverse and articulate individuals with well-developed senses of humor.

This observation, of course, contradicts the stereotype of coolly logical scientists or engineers who express their thoughts only in dry mono- or five-syllable words. But the stereotype could hardly be further from the truth. This point was again driven home at the American Petroleum Institute's recent refining meeting in Toronto.

The witty after-dinner repartee between competitive catalyst vendors, for example, or the colorful war stories of start-up specialists can be hilarious — and still packed with technical expertise.

The jesting touch can even crop up in technical presentations.

A case in point — one paper on instrumentation and control at Toronto began with this dismaying opener:

"I<2>L bipolar technology, even with its emitter-collector and multilevel interconnection problems, probably will replace N-MOS LSI in spite of the trend to silicon-on-sapphire insulating substrates."

But the author, F. J. Berto, a senior engineer with Standard Oil Co. of California in San Francisco, quickly got his nervous audience, including one editor with sweaty palms, off the hook with the statement:

"Before you all decide to break for coffee, this is not what my talk is about. But your company better have people on the payroll who understand this kind of argot."

The audience, of course, didn't break for coffee, because Berto's presentation, "The Impact of the Microcomputer on the User," proved to be a valuable and refreshing view of a complex field.

Unfortunately, for some mechanical reasons and for fear of doing some bodily harm by abstracting, Berto's presentation doesn't appear in this issue. But persons dedicated to instrumentation and control definitely should get a copy of this paper for themselves and for their associates who are not dedicated.

Not surprisingly, Berto is not your stereotypical engineer. He is a contributing editor for "Bicycling" magazine and has some fascinating thoughts on frame design, biking etiquette vis-à-vis motorists, and the lessons that bicycling holds for energy conservation worldwide.

He is a working editor, too. He bicycles 3,000 miles/year.

By the way, computer aficionados are indeed now talking about I<2>L bipolar technology, Berto says. He not only understands that arcane subject. He can communicate it, thanks to a finely honed sense of humor.

Saving the system

Howard M. Wilson **May 29, 1978**

If you want to know how, and how not, to run a trade association for a bunch of oil companies, Harry Morrison is just the man to straighten you out.

Harry Morrison has written a book.

He's given it the provocative title "Millions Are Plotting to Save the System," and it's just been published in Los Angeles by Lundberg Co. — the people who put out the Lundberg Survey for oil marketers.

Harry was the manager of the Western Oil & Gas Association during the years when California politicians were learning that it was politically profitable to turn the oil industry into a whipping boy.

It was WOGA's (and Harry's) job to explain to a skeptical public why hydrocarbons were polluting such sacrosanct environs as the lovely hillsides of Forest Lawn Cemetery, the beaches of Santa Barbara, or the atmosphere of sunny southern California.

Not all of WOGA's problems were with the public. Its own members sometimes got sore and quit because they couldn't support a position taken by the organization.

But they usually rejoined.

One hard-headed lawyer for a major oil company, according to Harry, explained trade associations this way: 1. They cannot be expected to do any good; 2. They are capable of doing great harm; and 3. There is no possibility of abolishing them.

While Harry doesn't agree with this lawyer that the way to keep trade associations in line is to keep them understaffed and underbudgeted, he does recognize that associations must reflect the view that the member companies are genuinely searching to preserve free enterprise in America while at the same time recognizing the need to work in harmony with government and with society as a whole.

Citing the business philosophy of several oil executives, Harry Morrison draws these conclusions:

"The oil industry leaders are saying big business is needed, public support is needed, and there must be a real and continuing and never-ending dialogue. There must be cooperation between business and government, there must be honesty in action, not just avowals in talk. And management must listen attentively to the public and learn to understand what the public wants. Particularly, it seems to me, they are saying all these things must be done to save for the people the best tool there is to accomplish the job — the privately owned corporation."

Turning then to the role of the trade association, Harry Morrison says these leaders and their company employees must use the association as an integral part of the "constantly changing, coordinated effort" to keep private industry alive.

These leaders in the private sector, together with their employees and their trade associations, are "the millions who are plotting to save the system."

Jug hustler, editor, awardee

Gene T. Kinney **June 12, 1978**

Our man McCaslin hasn't always been a chronicler of the world's wildcats and wildcatters.

It only seems so to the past generation of Journal readers, who have watched the continuing saga of exploration roll off his typewriter for 27 years. Before he began writing about oil finders, Mac used to hunt places for them to drill.

As a green geologist right out of the University of Oklahoma, he roamed the outcrops around Midland for a Tulsa-based seismic firm.

"I was hustling jugs in the late 1940's," he recalls, "before Snyder, Tex., became Kelly-Snyder field."

Before and since that time, some of the world's great exploration thinkers have practiced their trade in that petroleum province. They have made the West Texas Geological Society one of the finest groups to be found anywhere.

Jack McCaslin has spent much of his Journal career highlighting their successes and failures, first as district editor, then geological editor, and finally exploration editor. Now, they've turned the tables on him.

The society has chosen Jack to receive its Distinguished Service Award for 1978. The award recognizes and honors individuals who have made exceptional contributions to the industry, the science, or profession of geology.

Mac's own experience spans the Mid-Continent, the Rockies, and Texas, to which he devoted most attention. But his recollections go first to the people who enriched that experience.

"I had the honor to work under people like Frank Gardner, whose name is and was synonymous with geology and exploration and discoveries," he says. "Got to work with men like Neil Williams who was the Journal's representative at the Leduc discovery for one, saw Oklahoma City become a field, and untold others."

Best of all, Mac has had the privilege of working with AAPG greats like Mike Halbouty, Lewis Weeks, Ira Cram, Fred Lahee, Hollis Hedberg, Morgan Davis, Sr., Dow Hamm, Wallace Pratt . . . name 'em.

He fondly recalls a lunch years ago with John C. Sproule, "the Canadian's Canadian, the geologist's geologist," who marveled: "I don't know how you and Frank can fill all those pages every week."

Wallace Pratt, whose impressions as a reader far predate the McCaslin era at the Journal, wonders how we cram so much in the allotted space. "You confine John McCaslin to less space, I think, than George Blardone, reporting weekly out of Tampico in 1916, occupied with his comment on Mexico alone," this venerable figure chided in a recent letter. "But in nearly every issue of the Journal, McCaslin's columns teach me a little more about geology in the search of oil."

Our hat is off to Mr. Pratt for that generous comment; to the West Texas Geological Society and its president, Thomas J. Hansen, for recognizing our exploration editor; and, of course, to our man McCaslin, for his distinguished service to us and our readers for 27 years.

The free market according to DOE

John L. Kennedy **July 10, 1978**

It had to happen. The Department of Energy has come up with a good — no, a great — idea.

It proposes to allow a commodity to sell at "prices determined by normal market forces without government intervention."

Yes, it's true. But that's not the whole story, so get hold of yourself.

As it turns out, the glow of this fantastically bold concept is slightly dimmed when you know what commodity will be subject to these "normal market forces."

Don't get your hopes up — it's not natural gas. It's not natural-gas liquids. And it's not crude.

Okay. Are you ready for this?

The DOE is going to allow a free market in gasoline ration coupons!

DOE has prepared a standby gasoline rationing plan to be used in case of a U.S. fuel crisis. Under the plan, DOE will give motorists permission to buy and sell ration rights in a "white market."

Kinda restores the old faith, doesn't it? Makes you sorry you said what you did about the Government's attitude toward free enterprise, right?

It also makes you wonder if maybe such a mechanism would work for, say, natural gas. If it would, it might even be possible to apply the same approach to a substance in the liquid state.

I don't have any particular liquid in mind. It's just a thought.

Of course, even under "normal market forces," all is not wine and roses. In the case of gasoline rationing, for instance, I'm sure you realize there will be some minor costs involved.

Energy Sec. James Schlesinger says it will take $100 million to get the plan off the ground and $180 million/year to keep it flying. That doesn't include the $26 million worth of ration coupons printed earlier. They will have to be destroyed because they were not serially numbered.

Here's some more good news. It will take only 5,000 persons to run the rationing program. Think how many it would take if DOE planned to "intervene" in the ration coupon market.

But back to the possibility of letting something like natural gas loose under normal market forces. As our special report in this issue indicates, there is a lot of natural gas in the world. Some of it is not near the market, but that's only a political and technological problem.

If we're lucky, we won't get a chance to test free-market theory with gasoline-ration coupons.

But maybe it would be worth a trial run on something not quite so critical — like oil or gas.

If it did work . . . well, who knows? Those ration coupons might never have to come out of the DOE vault.

OPEC and the press

Roger Vielvoye **December 25, 1978**

No event on the oil industry calendar brings journalists flocking around like a price fixing meeting of the Organization of Petroleum Exporting Countries.

For a key meeting like the one in Abu Dhabi, reporters outnumber delegates two to one.

Most newsmen want to know the results before the meeting has even started. Unfortunately, this desire for interviews and quotable comments conflicts directly with the tight security net the host government throws around its guests.

It was not always so. Before the kidnapping of ministers in Vienna in 1975, an OPEC meeting was a relaxed, informal affair. Ministers and the few newsmen who attended in those days could mingle in hotel lobbies. Information flowed freely.

Contrast with the Abu Dhabi meeting could not have been greater. Six days before the meeting was to begin, the seafront Abu Dhabi Hilton began to clear out its regular guests. Then the security men took a hand. Everyone — staff, officials, and press — had to have two security passes, one for the meeting of the Organization of Arab Petroleum Exporting Countries (Oapec) and a second for the OPEC gathering.

By the evening the guests are all gone, and the army has moved in. Armored cars surround the Hilton. There are open land rovers with rear-mounted machine guns everywhere.

By the time the first delegates start flying in for the Oapec meeting, an airport-style baggage screening device is installed in the lobby.

By OPEC standards the Oapec meeting is a low-key affair. The press ranks are thin. But it's the lull before the storm. As the non-Arab delegates start arriving at the airport, so does the international press.

Two days before the start, delegates begin meetings with the press despite the security arrangements.

Saturday the OPEC conference opens with a formal session and quickly breaks up into more casual meetings. The argument in the afternoon centers on whether Arabic should be the official language of OPEC. Everyone is playing it cool on prices.

Journalists are still separated from the delegates by a security screen. But by evening word begins to filter through that the price increase is agreed and will be phased over four quarters. Only the mechanism remains to be decided.

Sunday the ministers meet to hammer out the final communique. By late afternoon the details are agreed, and retiring Sec. Gen. Ali Jaidah calls the first press conference to announce the decision.

Abu Dhabi Oil Minister Mana Saeed Al Otaiba also meets the press, but it's Yamani that most are waiting for. Yamani says he is not happy. He is subdued in his response to questions.

Then comes the mad rush to penetrate the security screen and meet as many ministers as possible. Most ministers have spent 4 or 5 days cooped up in the hotel and don't want to prolong the experience. Neither do the journalists.

By next morning most of the OPEC officials have left.

Next venue: Probably Geneva in June. But then there won't be the pricing issue to draw the crowds.

Saudis add another Mexico

Bob Enright **January 1, 1979**

Making accurate calculations of reserves in any given oil or gas field is difficult.

The reservoir engineer early on has only well tests and knowledge of reservoir rock properties in the immediate area of the well bore. Later on he has production performance, but it is still impossible to pin down how much rock porosity and permeability fluctuate between wells.

It is even more difficult, thus, to accurately fix reserves in an entire country with hundreds or thousands of reservoirs. The larger the area, the more the margin for error.

Having said this, however, we submit that the wild reporting of so-called Mexican "reserves" in recent months does great violence to even the most liberal margin of uncertainty.

Guilty of tremendous exaggerations of the size of Mexican reserves have been the press (which can perhaps be excused) and U.S. government energy leaders and oil-company officials (which can't). Oil-company types who have picked up and repeated exaggerated "reserves" figures, to be sure, have had marketing, refining, or such backgrounds. They weren't production oriented. But the public doesn't differentiate. He is an oilman, and who should know if he doesn't?

So let's set the record straight. In the case of Mexico, proved, probable, and possible reserves; oil-in-place figures; and even "potential reserves," which equate with undiscovered potential, have all been thrown into the pot. This figure then has been compared with actual proved reserves of such giant producing countries as Saudi Arabia.

It has been proclaimed in several articles we've seen that Mexican oil "reserves" are 200 billion bbl or "as much petroleum as in Saudi Arabia," as one stated. Thus, Mexico suddenly became another Saudi Arabia.

That 200-billion-bbl figure, friends, is Petroleos Mexicanos' estimate of Mexico's oil, gas, and gas liquids potential — not its estimate of proved oil reserves. You might as well say that U.S. reserves are 200 billion bbl. That would include some 28.5 billion in proved reserves, some billions more of probable reserves, and up to 130 billion bbl of potential oil. The latter figure is the upper end of the latest "undiscovered recoverable oil" estimate for the U.S. by the U.S. Geological Survey. The U.S. total would be far higher still if you threw in the equivalent of proved, probable, and potential gas and gas liquids.

Actually, Pemex's latest figure (as of Jan. 1, 1978) for Mexico's proved reserves is 20 billion bbl. Even this figure, when on occasion correctly quoted, is misunderstood. It also includes the liquid equivalent in barrels of the country's gas and gas-liquids reserves. Mexico's proved reserves of oil as of that date were about 14 billion bbl.

That does not, as some have said, put Mexico in a class with Saudi Arabia where proved oil reserves alone are pegged by Petromin at 165.7 billion bbl. OGJ estimate of Mexican proved oil reserves as of Jan. 1, 1979, is 16 billion bbl. That, incidentally, is about the amount by which Saudi Arabia increased its reserves from the year-earlier estimate.

Does this mean Saudi Arabia had added another Mexico?

Let us spray

John L. Kennedy **February 12, 1979**

In mid-1977, I implied that consumers, taxpayers, and the industry would get nothing for the $10.6 billion the Department of Energy budgeted for its first year's expenses.

That's not entirely true. We are getting something, and I always like to give credit where it's due.

The awful danger of having plenty of oil and gas, for example, is no longer a serious threat.

Thanks, DOE.

And DOE is not relaxing its vigil, no sir! For instance, it is now battling another evil that threatens our economy: hydrocarbon aerosol propellant. You know, spray cans — shaving cream, wax, insect repellant, deodorant, perfume, antiperspirant.

Now that's what a national energy policy is for! Incentives for increased oil and gas production will take care of themselves. It is overpriced protection against wetness that could destroy our way of life.

Ever watchful, DOE decided last year it could regulate hydrocarbon aerosol propellants (HAPs) under its authority to allocate and set prices for "crude oil, residual fuel oil, and each refined product."

But then some cheeky manufacturer of HAPs filed — and was granted — a motion for a preliminary injunction against DOE that prevents DOE from regulating the price of HAPs until the case can come to trial.

The manufacturer says: (a)HAP is a specialty chemical, and therefore it is exempt from price and allocation rules and (b) besides, it tried several times to get the former Federal Energy Administration to clarify the rules and FEA did not disagree with the company's position described in (a).

This firm, when fluorocarbons were outlawed as a propellant, developed a substitute and convinced the Environmental Protection Agency it was safe.

But if it has to comply with DOE pricing rules retroactive to 1973, it would have to refund about $10 million — three times the stockholders' equity. If it doesn't refund and is in violation, the penalty could amount to $20 million. Each violation can get you up to a $10,000 fine, and each day the violation continues it's another 10 big ones.

Having to do either could kill the company. The regulation would live to fight again.

But this is a serious offense! We are talking about maybe 4,000 b/d of propane and butanes used to make HAP. That's almost 1% of U.S. production of these products.

If DOE can get the court to call HAP "product" instead of a "specialty chemical" it can concentrate on the less critical details of our national energy plan — a promise to deregulate gasoline, for example or "regulatory reform."

In the Nov. 13 issue of The Journal, DOE's O'Leary said, "We really want to get out of the regulations business."

Unquote.

Who said bureaucrats don't have a sense of humor?

The terms of technology

W. D. Moore III **March 5, 1979**

Only someone outside the industry would realize what a specialized technical vocabulary we employ in drilling.

Not only are the words unique to the oil patch, but within it they may have more than one meaning. And sometimes, the second meaning is better known than the first.

Here are definitions begged, borrowed, or stolen for some of those ambiguous terms.

Hard hat — A device which reduces choke pressure to an acceptable level. Device is placed over the pressure gauge.

Geolograph — A recorder which works automatically while drilling but requires an operator while pulling on stuck pipe.

36 — A tool for connecting 1/4-in. copper tubing to a water cooler and for moving pipe around by the threads.

Pit drill — A game in which the mudman tries to rewire the warning horns while the driller isn't looking.

Test BOP's — An oilfield term for coffee break, used mostly on the morning report.

Slicker suit — Sometimes used to protect weight indicator while jarring or running a long string.

Fill the hole on trips — An obsolete practice used before drillers learned that the hole would usually fill itself.

Morning tour — Offshore, this is the time for cleaning the rig and removing any excess mud. In hard rock country, morning tour means a new rig configuration where the assistant driller works between No. 1 and No. 2 engines and the driller and remaining crew are split between the mud room and the logging trailer.

Wiper trip — An invention of iron-clad Company men to keep the crew in motion during a long bit run. Below 8,000 ft a trip is called a journey. Below 15,000 it's called a #!!%# pilgrimage.

"That boat oughta be there anytime now." — Generally, this doesn't mean anything.

Cement — A carrier for additives.

Mud — See above.

Pop-off valve — An emergency relief valve used to rid the rig of visitors.

In memoriam — Doc Nelson

John L. Kennedy **April 23, 1979**

We have sad news to report.

Dr. W. L. Nelson, a member of the Journal family for more than 42 years and an expert on refinery processes and costs, died Apr. 14 in Tulsa.

Dr. Nelson began providing Journal readers with technical and cost data on refineries in late 1936.

The refinery construction cost index he developed first appeared in the Dec. 15, 1949, issue of the Journal. For almost 30 years, it served as a benchmark — a place to start — for refinery managers and planners.

After developing the Nelson index, he continued to supplement this widely accepted estimating and planning tool with comprehensive articles and a long-running series, "Questions on Technology."

These features, spanning 42 years and more than 2,100 issues, provided an enormous amount of information for the Journal to pass on to its readers.

But the number of pages was not the significant feature of Dr. Nelson's contribution to the refining industry's development.

It was the quality of the material.

The Nelson cost index has been the index for refiners for three decades. Readers often call asking for yet-unpublished indexes. They need the latest figures and they don't want to wait another 2-3 weeks to read them in the Journal.

Dr. Nelson's indexes and his articles were so valuable and timely because of his day-to-day involvement in the refining industry.

His active role as a consultant in refinery design and operation made his writings up-to-date and credible. He served as consultant and adviser to national governments and to private companies throughout his career.

He also headed the department of petroleum refining engineering at the University of Tulsa for more than 20 years and still found time to author two books and several manuals.

Because of his interest in serving the refining industry and his careful planning, we will be able to continue to publish the monthly Nelson cost index and the Quarterly Costimating series.

And Dr. Nelson had a number of important Questions on Technology "in the bank" which we will be publishing during the next several months.

We will miss Dr. Nelson at the Journal. And our readers will miss his insight into refinery operations.

But the industry will continue to benefit — directly and indirectly — from the knowledge he shared.

Life at the front

Howard M. Wilson **May 14, 1979**

Journal readers who live in cities where oil is the king bee industry — places like Houston, Tulsa, and New Orleans — should take a trip to combat country to learn how their less fortunate colleagues are faring.

Specifically, we are talking about the battlefront in Los Angeles.

Our identity as a reporter for an oil magazine is, of course, well known among our friends in suburban Glendale, where the automobile is not just a way of life; it is life itself.

After waiting a couple of hours in a line to top off a tank with 4 gal of gasoline, our friends are demanding explanations.

Since we're the fount of all wisdom on this subject, we catch the demands. With sweet reasonableness we trot out our old standby arguments — the Iranian shortfall, lack of refining capacity (occasioned by environmental constraints), sulfur content of Prudhoe Bay oil, soaring demand, rebuilding of fuel oil stocks, price controls, strangling regulations, and lack of drill sites.

With condescending smiles, but clearly with lack of comprehension or agreement, our friends are not convinced. They think they destroy all our arguments with an irrelevant body blow: the hefty earnings reports of oil companies.

Pained, we trot out our defenses again — the comparison of oil's earnings with other industries, the bad years vs. the good years, the need for profits to find more reserves.

When this fails, we play our trump card. How, we ask (this time with something less than sweet reasonableness), do you get to work each day? How many miles does your gas guzzler get to the gallon? How many persons ride in the car with you?

With these questions answered defensively, we explain that for 9 years we have driven 1 mile to the bus stop and ridden the remaining 9 miles to the office via public transportation. We save money and we save gasoline.

We must, however, make this an honest report. Our trump card often is overtrumped. Most people in the Los Angeles area can't get to work by bus — even if they want to.

But more are trying.

The proof? On the Monday that followed the summer's first panic-button weekend, here's what happened to us:

After we had sailed by our Mobil station, where two blocks of cars were waiting for service, we parked in the public lot next door, boarded our bus, found standing room only for the trip to the Journal office. It was our first stand-up ride since the Arab oil embargo 5 years ago.

Conservation and chicken soup

Jim West **June 18, 1979**

No doubt about it, energy is becoming more precious every day.

Industry has long known that, in view of its rising cost to find, produce, transport, process, and market oil and gas.

Now, with sharp increases in the price consumers pay, the energy conservation message is getting through to the U.S. public.

U.S. motorists, for example, have cut their driving from levels earlier in the year as a result of sharply rising prices and tight supplies of gasoline. They are following the lead of conservation progress in the commercial and residential energy sectors, where efforts are building to cut utility bills for space heating and air conditioning.

The American Petroleum Institute pinpoints one segment of industry's conservation program.

An API survey unveiled just a few weeks ago found that a program by U.S. refiners had slashed the amount of energy required to process a barrel of crude by 19.6%, compared with the base year 1972. The cutback in energy consumption, API reports, is equal to a savings of about 330,000 b/d of crude. And refiners nearly have achieved a targeted 20% reduction in energy consumption by 1980, compared with the base year.

Consumer efforts are evident in cities such as Chicago. There eight companies and organizations won conservation awards from the Use Energy Wisely Committee of Greater Chicago. The awards recognized efforts that cut consumption of electricity and natural gas.

On the West Coast, San Diego Gas & Electric Co. says that more than 2,000 San Diego County homes and businesses are saving energy by installation of a unique sun shading system — solar shade screens — purchased from the company and installed by licensed contractors. Commercial and residential sales of the screen amounted to 700 in 1977 and more than 1,600 in 1978. SDG&E expects another increase this year in sales of the screens, which tests show can cut a typical home's air conditioning costs by up to 32%.

In Big Spring, Tex., Cosden Oil & Chemical Co. reports that crude converted to foamed polystyrene is cutting consumption of electricity produced from oil, gas, and coal. The price of fossil fuels has reached the level at which it is cost effective to incur the costs of converting crude to aromatic chemicals to ethylbenzene to stryene monomer to polystyrene to foam sheathing board in order to insulate a home using electricity for heating and cooling. Energy use has been reduced by 23% in a test home, Cosden reports.

The congressional Office of Technology Assessment figures that readily available methods to conserve energy in home heating and cooling can yield a cumulative savings equal to 13 billion bbl of oil by 2000 — with no change in lifestyle or comfort.

No realist believes the U.S. can cure its energy ills by conservation alone. But conservation, like grandma's chicken soup administered to ease the flu, at least couldn't hurt. And saving energy makes good economic sense at today's prices.

Middle East exports and politics

Roger Vielvoye **July 9, 1979**

Since our last Middle East Report, the area has undergone fundamental change.

Ousting the shah and attempts by the Ayatollah to introduce a more austere, inward-looking Iran has, as everyone knows, caused worldwide oil shortages and sent ripples of political uncertainty around the Persian Gulf.

Throughout the Middle East, rulers, whether monarchs or presidents, are looking nervously over their shoulders for the first sign of trouble from political or religious sects or some ethnic minorities that abound throughout the region.

But outside their own countries, leaders of the oil exporting countries move and talk with greater confidence than before. The shortfall in supplies from Iran has done what successive meetings of the Organization of Petroleum Exporting Countries failed to accomplish during the years of oil surplus. It has increased the price of oil in real terms.

OPEC and the Middle East oil exporters in particular are at the center of the stage. The fluctuations of Iranian exports and Saudi Arabia's internal debate over production ceilings command the attention of oil buyers everywhere.

Swinging price increases are accompanied by lectures to the industrialized countries about their profligate consumption of energy. For countries that previously have done all the lecturing it's galling, especially because there's no denying the truth of the message that conservation is vital and only higher oil prices will provide the incentives to develop alternate energy sources.

Ironically, the consuming countries make less fuss over the oil exporters' criticisms of help for the Third World even though OPEC members are on less firm ground.

Ominously, from the consumers' viewpoint, this new-found confidence could soon spread into more than just talk. Oil shortages give the "oil weapon" a new edge. A number of countries are making thinly veiled threats that it could be used again.

Iraq says it will halt supplies to Canada if it moves its embassy in Israel from Tel Aviv to Jerusalem. Libya threatens similar action against the U.S. if the transport planes ordered don't come through. Nigeria says it will move against any country recognizing Rhodesia and may penalize BP if it sells North Sea crude to South Africa.

So far, they are threats in pursuit of individual political objectives. But time may be near when Middle East exporters once again talk of using the oil weapon collectively in pursuit of their one common goal: establishment of a Palestinian homeland and withdrawal of the Israelis from the West Bank.

Frustration at the lack of progress toward this goal is growing throughout the Arab world. Even the Saudis have warned the U.S. of the possible consequences for oil supplies of failing to hustle Israel into a Palestinian settlement.

Our next Middle East Report could be published in even more depressing circumstances. For no matter how much time and money is poured into insuring that crude is physically available, the level of exports remains firmly a political decision.

The clod lobby

L.R. Aalund **July 23, 1979**

Evidence indicates that the U.S. clod lobby has already shaken Saudi Arabia.

"Clod," I hasten to add, is primarily used herein in the sense of "soil," a symbol for like thinkers, similar to others such as "Clamshell," "Roadrunner," etc.

The clod lobby is that loose coalition of some farmers and their allies in universities, state houses, and Washington who are preaching to the public that agriculture can pull the U.S. out of its energy misery. The solution is fuel from fermented grain, sugar cane, and banana peelings — gasohol, in short.

The evidence of the coalition's first victory over the oil cartel is circumstantial. But who can deny that shortly after announcement of only a few gasohol projects this spring, Saudi Arabia suddenly increased its crude oil production 1 million b/d?

The Organization of Petroleum Exporting Countries knows that the U.S. has a huge resource base. If the U.S.'s entire production of corn, wheat, barley, oats, and grain sorghum is fermented, it will produce an amount of 200-proof alcohol equivalent to about 1 million b/d of gasoline (one-seventh of U.S. demand). Throwing all U.S. sugar in the still would boost the volume more — not enough to cut the cord with the Persian Gulf but enough to put some slack in it.

And OPEC, after watching its coffers fill, indirectly as a result of the U.S. gasoline appetite, can't totally reject the idea that Americans would rather drive than eat bread.

Americans would obviously have to make some sacrifices for a total gasohol commitment.

The price of bread would soar, maybe up to $3-4/loaf. There also could be odd and even day bread lines. A painless method of attaching a license plate to each buyer would have to be worked out.

A positive development would be that the heat would be off the oil industry. But there would still be nasty little stories about full grain elevators in Olathe, Kan., as the shortage spreads; or about a freighter laying off Galveston, riding high in the water, proof that she's loaded with a cargo of Puffed Wheat and waiting for the price to go up.

Some fringe elements in the clod lobby also are advocating another tactic: unsheathing the wheat weapon — "no oil, no wheat." This would require the founding of a grain exporting cartel with other countries. It probably would be dubbed Organization of Grain Exporting Countries (OGEC). That would be a long, hard road.

Also those innocents don't realize that to break OPEC the new OGEC would have to break Saudi Arabia, which may own Kansas by then. They don't realize that the mere 6-7 million people in the desert kingdom aren't locked in to a breakfast with Wheaties, biscuits, and grits.

Finally, a controversy would arise over how much grain goes to gasohol and how much for the wheat weapon. With congressional guidance, the Department of Agriculture surely would foul up the works with an entitlement program.

This one came tough

Robert J. Enright **August 20, 1979**

Of the 52 issues cranked out by the Journal this year, a startling 32 will be special issues or contain special reports, such as that on Mexico in this issue.

That is an OGJ record. Our normal quota in recent years has been 28 or 29. But this is the year of the World Petroleum Congress and the International Petroleum Exposition. We didn't want to cut worthy geographic, statistical, and technological specials. So we added on instead.

It figures, when you run this many specials, that you'll likely have problems with some of them due to unforeseen political, operational, or other circumstances. And the unforeseen indeed did occur with this one.

Prime responsibility for the Mexico Report was given to Ted Wett, OGJ petrochemical/gas processing editor, and Shannon Matheny Jr., production editor. Ted and Shannon flew to Mexico City in June to begin the big info-gathering job — and arrived on the very eve of the blowout of the 1 Ixtoc in the Gulf of Campeche. The Ixtoc threw a formidable monkey wrench in Shannon's initial plan for assembling information on Mexico's surging exploration, drilling, and production programs. The direct sources for this data, top Pemex E&P operating personnel, departed post-haste for the blowout site and were totally preoccupied with shut-in plans and efforts. They had little time for anything else.

We don't have the space here to recount how Shannon fulfilled his mission. But we think you'll agree, when you've read this special, that the job he did under such circumstances is remarkable.

Coupled with Ted's thorough coverage of where Pemex stands in its intensive oil and gas processing buildup, we'll wager you won't find anywhere else this kind of across-the-board coverage of Mexico's drive to become one of the world's foremost oil powers.

There's another reader bonus in this report — a tear-out wall map of Mexican oil and gas fields, plants, and pipelines — which greatly excites our foremost in-house expert on Mexico. That is Gus Pena, publisher of our sister publication, Petroleo International. Up to now, Gus says, there has been no such map available anywhere — even in Mexico.

There is one thing you won't find in this report: an article on the Ixtoc blowout and spill itself. That's because there's little to report on this spectacular wild well that we haven't already said or are saying in our weekly coverage in the Journal news section.

Pemex, as it always has in OGJ coverage of the Mexican oil scene, cooperated nobly in our Mexico Report effort, despite the disaster off its Yucatan coast. We thank them for it.

The inscrutable Americans

Gene T. Kinney **October 15, 1979**

When the Journal editor travels abroad, he runs into a barrage of questions from oilmen baffled by American energy paralysis.

They are puzzled by the contradictions of public policies, and by the official and public attitudes toward oil companies and prices. As one who writes often on the subject, the editor is called upon to explain. If an American isn't careful, he will spend more time being interviewed than getting answers to his own questions.

We encountered this repeatedly during the World Petroleum Congress in Bucharest and in later calls in Paris and London.

A reader from Belgrade, spotting our name tag during WPC, was bemused by American complaints over $1/gal gasoline, less than one-third the price paid in his country. Commenting on the paradox of excessive consumption and price controls, he said: "Yugoslavians will gladly trade our prices for yours."

In western Europe, as in Japan, oilmen and government officials are anxious about American production and consumption trends. They consider the U.S. a part of the energy problem on a par with OPEC.

Will America be able to stay within the imports ceiling agreed to at the Tokyo economic summit? Or will America continue to consume a rising volume of world oil supply, pushing prices even higher than the official OPEC schedule? Why doesn't the U.S. stabilize the dollar so that OPEC isn't encouraged to raise prices so often to compensate for its decline?

We get the same questions at home from foreign visitors, for whom Oil & Gas Journal is a regular stop on their tours of the oil patch. In the midst of a crude shortage, there is a surplus on the West Coast. Why don't you build a pipeline inland? America is the Saudi Arabia of coal. Why don't you use that position as your ticket to energy independence?

Oil companies must be successful if America is to solve its energy problem. Why do politicians want to break them up? Unlike most countries, you are blessed with abundant energy resources. Why don't you develop them?

We tried to answer these questions posed recently by a delegation from Nigeria, Zambia, Mexico, Canada, Poland, Romania, and Pakistan. We recited once more the litany of price controls, leasing restrictions, and environmental rules which frustrate energy projects of all kinds. That explanation only brought another question.

"If the causes of your energy problem are so clear," said our friend from Pemex, "why don't you eliminate them?"

Why indeed?

It was a very good question. And we ran out of answers.

A lesson on the Washington press

Randy Sumpter **December 3, 1979**

There's nothing like getting the old bum's rush from one's peers to set a man's teeth on edge.

I ought to know. It happened to me recently.

I had traveled to Washington, D.C., to help Pat Crow, the Journal's congressional editor, cover a busy week of subcommittee hearings and floor action. I had a particular interest in one of the hearings — the House interior and insular affairs committee's deliberations on the proposed Coal Slurry Pipeline Act of 1979. So interested, in fact, that I showed up at the committee room 30 min early and sat down at one of the two tables plainly marked "press."

By the time the witnesses, most committee members, and some of the press had arrived, the room had filled nearly to capacity.

I was congratulating myself for being smart enough to come early and get such a great view of what promised to be a long and arduous hearing.

Was I ever wrong!

About 5 min before the chairman gaveled the meeting to order, a young woman walked over to my table, rapped the side of it with a stenographer's pad, and whispered, "Are you with the press?"

"Yes, I am," I said.

"Could I see some identification?"

I produced a business card. She fingered the edges of it for a moment and then asked, "Are you a member of the press accredited to the House Gallery?"

"Nope."

"Well, you'll have to move because these seats are for accredited members of the press."

It was my turn. I started pumping the questions in my best reportorial style. I didn't intend to be flimflammed out of a ringside seat.

"Are you a member of the committee staff?" I asked.

"No, I'm not."

"Well who are you with?"

I got an unsatisfactory answer along with the general impression that if I didn't move, my own fourth estate would supply the necessary manpower to move me. About that time, the chairman banged down the gavel. I got up and moved to the back of the room. Let me tell you what I saw during the next 6 hr:

- Two nearly empty press tables.
- Two members of the press trading jokes loudly enough to disturb others in the room.
- One member of the press doodling on committee-supplied notepaper.
- No members of the press rushing out the door behind a witness to ask additional questions.

The press. Ha!

What waiting on weather means

Roger Vielvoye **December 10, 1979**

"Waiting on weather" is one of those phrases people working in the North Sea use to describe the frequent delays that bedevil operations in Northwest European waters.

To anyone writing about the North Sea the expression is heard so often it is easy to lose sight of what the waiting can mean to construction workers and operation crews.

Under normal circumstances a severe North Sea gale blows itself out in a couple of days. But recently the whole area suffered some of its most prolonged bad weather for many years.

According to Imcos Marine, the privately owned weather forecasters with a top reputation among offshore operators, no records were broken in terms of the much-talked-about 100-year wave or 100-knot winds.

Instead there were 8 solid days of southerly gales with waves rising to nonrecord heights of 53 ft and winds gusting to nonrecord peaks of 79 mph. In between the seas rarely dropped below 15-18 ft, which makes the storm twice the length of the longest gale previously recorded in the Brent area. That's the kind of nonrecord that makes shore-based observers glad to be languishing behind a desk.

Out in Shell-Esso's Brent field, weather halted tanker loading for a couple of weeks — a major interruption in cash flow. But it wasn't just a shock to the balance sheet. Supply boats couldn't keep the platforms provisioned with fresh food, equipment and helicopter fuel, the shortage of which disrupted crew changes.

In the end Shell had to send the semisubmersible diving support vessel Uncle John from Brent to the Shetlands for supplies, proving the worth of semis in these conditions.

But even some of the semis ran into problems. Northwest of the Shetlands, BP's Sea Conquest dragged an anchor, a fate that also befell the Ocean Voyager, only this time the anchor dragged into the pipeline serving Occidental's Piper field.

Luckily, there was only superficial damage. But the field had to shut down for 36 hr while a check was made.

All in all, it was an eventful few hours for the Piper crew. Four children who had won an offshore trip as a prize on a U.K. television show were on board to watch the drama of the dragging anchor and were still on hand to see a fishing boat sink within sight of the platform. The crew were rescued by Piper's safety boat.

Even for the experienced crew it was quite a day. And perhaps the broadcasting of the incidents in a raging storm will help those on shore who benefit from Britain's new oil wealth appreciate the conditions under which it is acquired.

The energy writers' disease

Bob Tippee **December 17, 1979**

There's a new disease plaguing energy writers these days.

It's called scenariosis, and it comes from reading too many energy supply/demand forecasts and similar studies.

Diagnosis is simple. Victims begin to precede everything they say or write with, "It all depends."

Incidence of scenariosis has increased along with growth in complexity of the energy business. That complexity has altered the ways industry's prophets present their views of the future, which energy writers always hurry to report.

In more innocent times, industry's great thinkers could trace historic trends, project them into the future, test some theories, write a forecast, and go out and get a beer.

Energy journalists, in turn, could skim the table of contents of one of these stories, report the conclusion, throw in some quotes from the summary, and go get a beer.

That was before governments decided they could handle energy business better than energy companies and before the world economy came to rest in the palms of a few oil ministers.

It also was before computers.

Now, industry's prophets avoid pinning their careers to firm, conclusive statements about the ways things will be in the future. There are too many things that can go wrong with forecasts.

So the prophets write groups of projections, each based on a different set of assumptions called "scenarios."

Each scenario breeds its own forecast, so a study done this way serves up a whole smorgasbord of conclusions. It's like a multiple choice test where all answers are correct.

Corporate chiefs, for whom these studies are produced, must find them useful. They can use them like playbills, matching what's happening in the world with the closest scenario to keep reasonably close track of what's due next.

But for journalists, these multitiered projections are a pestilence. A writer now has to read all of a study at least once to make any sense of it.

This causes scenariosis.

Erstwhile stars of the profession, who once could dash off a week's quota of copy between hands in an all-night poker game, now spend hours trying to write punchy, focused lead sentences on studies with 8 or 10 or 15 conclusions and still present the qualifications accurately. Their thoughts get blurred. Instead of going for a beer, they quaff Perrier water at their desks.

Some give up altogether and become editors.

Does this mean there should be a change? Is scenariosis serious enough to warrant a revision to the simple statement, the unqualified assertion, the single-conclusion study?

It all depends.

OPEC and the Orinoco

Roger Vielvoye **January 7, 1980**

One of the alleged pleasures of being international editor of Oil & Gas Journal is foreign travel.

But spending tedious hours in uncomfortable airline seats to visit exotic Latin America and then spending even more hours sitting in an equally uncomfortable hotel lobby waiting for OPEC ministers to agree last month that they couldn't agree on world crude prices is stretching the definition of pleasure a little too far.

So after 4 days of confusion, boredom, and worry about deadlines, the airline seat that will carry you home from Caracas suddenly looks inviting. And if you can avoid the hazards of sitting close to children or beside someone with eight pieces of carry-on baggage, the return journey can be moderately comfortable.

Imagine then the horror at discovering that your seat on a jam-packed Air France 747 is next to a bouncing, roundfaced Venezuelan girl who, the moment you sit down, grabs your arm and chatters away in a language you've never heard.

The journey has all the hallmarks of a disaster. Instead, it makes up for all the frustrations of the previous days, a late arrival and missed connection in Paris, and lost baggage.

My traveling companion from Caracas back to London was aged 6, named Belkys. A few months previously she had been living with the Piaroa Indians on the Parguaza tributary of the Orinoco River. It's a remote, wild area of southern Venezuela where contact with the outside world is rare.

Belkys was an orphan when she was found and adopted by a French diplomat. In the intervening months, she had picked up a command of Spanish and was starting to learn French.

She made her first acquaintance with big city life in Caracas during the 15 days before our flight — too short a time to blunt the innocence of a newcomer to the delights of civilization.

Flying over the Atlantic Ocean was thrilling for her. Music from the headset produced amazement. And the first taste of cheese from the airline turned out to be an eyes-shut, shuddering experience.

As our flight went on, I was treated to drawings of the vultures that circle the Venezuelan interior and the plants and bushes surrounding the river that provides the Piaroa with most of their food. Generally the tribe has survived in good shape. But survival requires isolation.

Further down the river, the other tribes have been seduced into throwing away their canoe paddles for outboard motors. To raise money for fuel, the tribe had to abandon their self-sufficient society and produce a small surplus of goods for trading. From then on it was the familiar story of a slide into the poverty that haunts so much of the Third World.

The upper reaches of the Orinoco is a market the global oil industry can afford to ignore for the moment. But it's doubtful whether traders in Japanese outboard motors will allow it to do so forever.

A memo to the news editor

Patrick Crow **January 14, 1980**

Dear Jim:

1. As you know, the second Georges Bank lease sale is scheduled for mid-1982. It's not too soon to start thinking about this one, especially in light of my experiences in covering the past sale.

2. The first sale was scheduled for January 1978 in New York City. It was canceled about the time my airplane hit a down draft over Ohio and popped peas into my lap. As you know, I'm not one to complain. But that futile midwinter trip also resulted in a personal tragedy: I lost my navy blue, wool stocking cap in the back seat of a Yellow Cab. My uncle had worn that cap through World War II. It had been a true friend to me for 15 winters. I came home with a cold.

3. You again assigned me to cover the sale last November in Providence, R.I. As you know, I'm not one to complain, but after I had waited 14 hr — mostly sitting in an unpadded chair — the sale was canceled. The most memorable thing about the trip was a $12 cab ride to the airport — about a buck a mile, by my reckoning.

4. The sale was rescheduled for December, and I was sent to Providence once more. This time the sale was held — after only a 3-hr delay. Now I'm not one to complain, but you heard that some crazies threw motor oil at us. Although bracketed by near misses, I wasn't hit by a single drop. I thought I was lucky until you told me the company would have paid the cleaning bill. I had been meaning for months to have that suit cleaned.

Also, attached to my expense statement is a receipt for a $15 cab ride to the airport. Somehow I found a cab driver even more greedy than the first.

He must have been a foreigner, because when I protested, he only mumbled one English phrase repeatedly: "It's on the meter."

5. It is a credit to your perseverance that the Journal was the only major oil publication represented at the Georges Bank sale. But as much as I hate to complain, I'm warning you now that I'll quit if you assign me to the 1982 sale.

I think you can glean my reasons from the summary above. Besides, I don't think covering this sale has been especially beneficial to my oil reporting career. I know that some office wags claim that it has been my oil reporting career.

And even if I can't find another oil reporting job, I figure I can knock down a couple of hundred dollars a day driving a cab in Providence.

It's obvious to me that cab drivers are the main beneficiaries of these sales anyway.

Covering the OCAW strike

Greg Clock **April 21, 1980**

Some friends in Port Arthur gave me a Texaco hardhat with the word SCAB printed on it when I left Southeast Texas in 1976. The gift was in recognition of my labor reporting in that area, where unions are strong and strikes and subsequent violence occur with unusual frequency.

I'm still covering strikes, but I don't need the hardhat now. Instead of walking up to the picket line for the Port Arthur News, I'm using the telephone to cover the U.S. refinery strike for the Journal.

It's safer in my current job, discussing strike issues (long distance) with company officials and the union spokesman in Denver, the mild-mannered but pointedly anticompany Jerry Archuleta (he coined "pornographic profits" last year).

Are there any similarities between the local strikes of the mid-1970s and the current walkout? Yes. The violence.

Many union leaders remind the workers that violence hurts the strike effort. But it continues in 1980. Roofing nails at the refinery gate. Rocks thrown at cars. Steel balls fired from slingshots. Baseball bats used to smash windshields. The bombing of a refinery pipeline in Pennsylvania. The shooting death of a supervisor in Louisiana.

Sure, there's a possibility someone interested in damaging the union's reputation might be responsible for, say, the pipeline bombing. But company videotapes at refinery gates show that strikers are involved in much of the violence.

Having lived in Port Arthur, I'm also aware of the feeling of some workers that a strike justifies a little violence and harassment. It's part of the strategy, part of the game.

And union leader H. G. McClain of Houston, as quoted this year in the Port Arthur News, said: "Not only are we dealing with a cold-blooded and haughty industry, we are dealing with a dishonest industry."

With that quote in hand, some workers will be more inclined to believe that "the company" is indeed trying to deprive them of their just desserts, just like the miserly nonunion textile company in the movie "Norma Rae." And with such a holy and somewhat romantic cause — "lowly worker vs. Big Oil" — violence is more likely to occur.

All this despite the fact that unskilled teenagers at refineries in Port Arthur can make more than $20,000/year if they work the graveyard shift.

The strike is in its fourth month at several refineries. It continues because the local unions want amnesty for workers who have been fired for alleged strike violence.

Striking for better wages, benefits, and working conditions is one thing. But keeping nonviolent workers out of a job in support of those who smash windshields is another.

As I recall, some of the guys were sort of sensitive about those violence stories in the mid-1970s. Maybe I'll dust off that hardhat after all.

Leaping lizards!

Howard M. Wilson **May 19, 1980**

Dam builders have the snail darter.

Oilmen have the blunt-nosed leopard lizard.

This lizard, with stripes across its back, is a 9-in. reptile that skitters about the desert sands of western Kern County, Calif., chewing up insects and sometimes feeding on other blunt-nosed leopard lizards.

The grayish-green cannibalistic creature is, like the snail darter and the whooping crane, an endangered species. At least he's on the official government list.There's some question whether he really is endangered. The Fish & Wildlife Service is trying to determine that matter — a project that, it develops, takes years to complete.

Meanwhile, oil companies that want to explore a large chunk of potentially productive Kern County territory have been forced to delay or cancel many drilling projects while wrestling with a variety of federal agencies with a variety of interpretations on just how the Endangered Species Act should be administered.

In fact, during one spell of the negotiations with one company, Congress failed to appropriate enforcement money for the act, and the Fish and Wildlife Service couldn't take action on a drilling application. Result: The company paid $6,000/day for a standby rig for nearly a week.

Up to now the government has held — in bureaucratic parlance — nine "consultations" with oil companies and its own overlapping agencies.

Three of these consultations determined that the lizard, at the drillsites under debate, wouldn't be in jeopardy. Four determined he would be in jeopardy. Three aren't decided.

In two of the four cases where drilling would jeopardize the lizard's habitat, the oil companies settled matters by taking a leaf from the air pollution regulations book. They spent money on trade-offs.

How do you trade-off to save a lizard? Well, it's like this. You clean up the lease — move some tanks and piping, abandon some old makeshift roads, and create more open space so the lizard will have more running room. He particularly likes to mate out in the open.

No one really knows whether this does any good, but it seems to be satisfying some of the Fish and Wildlife people.

It's the farmers, not the oilmen, who created the lizard shortage in the first place. They've been plowing up the land — tearing up the gopher holes and mole holes where the blunt-nosed leopard lizard likes to lodge.

There is some indication that the lizard, as a matter of living proof, likes oil people. If you drive out to the entry gate of the Damson Oil Co. lease, about 7 miles southeast of Maricopa, you'll very likely find one of the little fellows sunning himself on the gatepost.

He's been living in the area for 3 years. His name is George.

If studies were oil

John L. Kennedy **August 4, 1980**

The $20-billion synfuels bill was enacted recently, and your Department of Energy is not waiting idly for the National Synthetic Fuels Corp. to go into business.

Nosiree! It's off to a running start toward energy independence with a program that apparently will make the feasibility study one of the main weapons in the U.S. energy war.

Of 110 proposed alternative fuels projects selected to be the first up to the trough, 99 are for feasibility studies. We'll get the whole package of 110 for a paltry $200 million.

Of the 99 studies, 40 concern the feasibility of making ethanol from various feedstocks. And 28 of the 40 will study the feasibility of making ethanol from corn, or corn plus something else — cull potatoes, molasses, whatever.

The feasibility study is a formidable weapon. But apparently it's trickier than a bureaucrat getting rid of his department's funds before budget time. So DOE is taking no chances. The feasibility of making ethanol from corn will be studied in 27 states!

You may still believe that nine women who become pregnant at the same time can't produce a baby in 1 month. Don't be too sure. A $12-billion budget and 20,000 employees may make a difference.

Besides, just because it's feasible to make ethanol from corn in Illinois doesn't mean it's feasible in Indiana. Or Kansas, or Nebraska. Or, for that matter, Massachusetts.

Further evidence of the fickle nature of the feasibility study is the fact that two are necessary in Indiana. Birch Bayh (D.-Ind.), head of the National Alcohol Fuels Commission, must be aware of special problems in the feasibility of producing ethanol from corn, even within a single state.

Or something like that.

From the proposed studies, one might also assume it is cheaper to study the feasibility of a large hypothetical plant than a small one. Only $99,700 is requested to study a 50-million-gal/year plant in Iowa, but more than $2 million is requested to study a 20-million-gal/year plant in Louisiana.

It's the old "economy of scale" rule at work, I guess.

Costs of the studies are listed as "requested from DOE" and footnoted with the comment, "to be negotiated." You may assume that DOE will negotiate these amounts downward. I'm not going to assume that, but you may.

But let's not haggle over details. If we don't use oil and gas for fuel, and the 28 studies bring huge advances in technology, a shining new age of corn power may even produce more energy than it consumes.

Swimming upstream, downstream

L.R. Aalund **August 25, 1980**

Elizabeth Jogeese, the able editorial assistant for the four editors in Oil & Gas Journal's Houston office, says the second week in June was "quite nice, almost like a vacation."

The reason is that three of her clients were "down south" that week rounding up material for the Latin American section in this issue.

Drilling Editor Bill Moore was so far south he experienced the chill of a new Argentine winter. Shannon Matheny, our production editor, was only a hop away from Houston in Mexico City, happy that the trip home wouldn't be as long as the one he made from the Persian Gulf in April. This writer, whose beat is refining, was in Venezuela.

It's not unusual at any given time to have a Journal editor a long way from his home base. But when I went to work for the magazine in the mid-1960s far less travel took place. The petroleum world could then be covered with an infrequent trip overseas and editors in major U.S. cities and London.

An editor could get an international story simply by getting in his car or on the subway or tube and going to visit a large independent or international major. They had most of the experts and their lines of communications stretched around the globe.

For operations, the U.S. Gulf Coast or California were the major arenas for big projects and applications of advanced technology by all divisions of the oil industry.

We certainly haven't forsaken these methods or regions. Local contacts are still invaluable, and the Gulf Coast is still a citadel of the oil industry.

But on the way to the 1980s, a restructuring of the petroleum world began. OPEC became a household word, and centers of influence and information cropped up in once unlikely places. Stavanger was no longer simply a name on a sardine can. To keep pace, the Journal increasingly sent editors overseas during the 1970s.

Another trend also is influencing our assignments. Though most of our writers are switch hitters who can handle stories on various industry branches, we don't, as a rule, send an editor whose forte is petrochemicals or refining into a region where the primary activity is drilling or production. I have never been sent to Dhahran or Aberdeen, nor has our drilling editor ever been to Munich on his expense account (Baytown either, he says).

But now, as Venezuela shows, a country once known mainly as a producer can become a worldwide leader in the employment of advanced refining technology. Mexico is pushing forward on the drilling/production and processing fronts simultaneously. Others are going upstream or downstream.

Either way, the Journal can go with them.

Scare-fare

Theodore Barrington **September 1, 1980**

I'll bet you didn't know that in the modern nobility of health regulators, Ralph Nader is known as Lord Airbag. He certainly exudes enough of it, and though his hissing has been quiet of late, one never knows when he'll huff again for that giant creampuff which is supposed to pop out from the dash and suffocate us in safety.

Frankly, I'm tired of safety. I'm weary of all the federal fear mongering and snobbery that goes with it — the regulatory King Pings who make a royal reputation out of unleaded gasoline, the seatbelt barons who leave me strapped of patience, and those uniformed, metronomic ladies-in-waiting on airplanes who recite the thrills of inhaling oxygen and shooting the rapids in mid-ocean. No one listens to them. Who can stand to listen to them?

And then there are those pill-bottle princes who have seen to it that not even Jack Lambert can take an aspirin without a pair of pliers.

I'll tell you, I'm weary of the whole thing. I'm tired of antibenzene queens and antinuke dukes and other such kook pretenders to the throne of what they hope will be the New Stone Age.

Truly, I wouldn't mind if they came in with some definitive evidence. But their superficiality, their scare-fare that appeals to modern me-ism, is appalling. I can see them 500 years ago, down at the dock as Columbus is getting ready to sail — crawling all over the Santa Maria, pawing for leaks, scowling at the unsanitary seamen, and finally — having gained access to the ship's charter — canceling the whole trip because Columbus doesn't have Flat Earth insurance.

If this sounds like hyperbole, ask anyone who has to deal with the Occupational Safety and Health Administration, or the Environmental Protection Agency. If these bureaus had been around when Lindbergh was getting ready to attempt the Paris flight he would never have gotten off the ground. And Ford would have had to junk his magnificent Tin Lizzie.

And what about the following nobleman — a real one? The naturalist Baron von Humboldt traveled for months in the 1800s, with one assistant, through an "impenetrable" Panamanian jungle — with no radio, no modern conveniences, no bruise-and-cut kit from OSHA. He collected 200 soil samples, made drawings of more than 400 species of animals and plants, and wrote several books on the journey which are still in print today.

I've been in that jungle, and I can assure you OSHA would have not approved of the baron's journey. EPA would have instantly declared the place off-limits.

But I guess if you speak of jungles, OSHA and EPA do have their noble pride. Their jungles are smaller and certainly less dangerous. Selling safety nostrums to the great unwashed must therefore be a full-time job. And I suppose scaring the natives helps.

Hurricane watch

Randy Sumpter **September 8, 1980**

It may be an old saw, but it still cuts true: Safety pays.

And when it comes to hurricanes, the Gulf Coast industry embraces that maxim without exception or delay. Each year about this time operators and producers shut in wells, secure rigs and platforms, and evacuate 5,000 or more workers from Gulf of Mexico installations in the face of hurricanes.

Regrettably, the storm preparations alone often cost lives.

Twice within the past year these expensive and complicated precautions have paid dividends — once when Hurricane Frederic waded ashore at Mobile, Ala., and recently when Hurricane Allen disrupted operations on the lower Texas coast. Both were glancing blows that never threatened the heart of the industry in the central gulf.

Even the moderate damage in both cases was enough to justify the preparations. Frederic worked over Chevron U.S.A. Inc.'s Pascagoula, Miss., refinery and damaged Mobil Corp.'s drilling operations in Mobile Bay. Allen exacted a higher toll: two jack ups sunk, a third damaged, and two drill barges swamped. The full inconvenience to shipping because of storm-related shoaling can't be calculated.

How much higher would the toll have been if the central gulf had been involved?

Unfortunately, we know. Eleven years ago to the day that I wrote my modest wrapup story on Hurricane Allen, another Journal Gulf Coast news editor wrote a story that said in part:

"At least two platforms, with their rigs, were destroyed, and many other rigs were damaged when the high-velocity, although somewhat narrow, hurricane slashed through the offshore area just east of the delta on the night of Aug. 17.

"Damage estimates of major operators indicated the total cost to the industry would be at least in the $50-million range and could rise to $100 million. This would be less than the cost of Betsy in 1965 and Hilda in 1964, both of which moved ashore on the west side of the delta."

The year was 1969 and the writer was Howard Wilson, now the Journal's West Coast editor. The story was headlined, "Oil installations hit hard by Camille." Howard's story ran almost two full Journal pages. The last page was devoted to an itemization of storm-related damage.

I'd rather not scoop Howard on that one.

Paul Reed — one of the giants

Gene T. Kinney **September 22, 1980**

Friends and admirers all over the world have their special recollections of Paul Reed.

We encountered him in all manner of places. Eventually we were no longer surprised to run into him at the ice rink in Rockefeller Center, on the ski slopes of New Hampshire, on a pipeline project in the jungles of Colombia or the lowlands of Europe, kneeling at St. John's, St. Patrick's, or St. Peter's, scuba diving off Jamaica or sailing off Marblehead. If we were old enough, we even knew him as one of that daring pioneer breed, the World War I pilot.

Paul Reed, the tall, patrician Harvard man, was all of these and much more. Most of all, friends knew him as our pipeline editor for 27 years before his retirement in 1962.

When he joined Oil & Gas Journal in 1935, the main strength and awkwardness period of pipelining was just yielding to the burst of technology. The tong gang and screw pipe were going out, and welded joints were coming in.

When Paul first wrote of construction, the wooden horse and mop pole had given way to the tripod for lowering in, and that to the modern sideboom tractor. Rule-of-thumb design was being replaced by precision engineering. Paul duly recorded the miracle of shipping multiple products in a single line.

All this fascinated the Journal's pipeline editor. But fascinating as pipelining was to Paul, he was even more taken by pipeliners. And they were taken by him.

Bill Heltzel and C. M. Scott . . . Oscar Wolfe and Ted Aude . . . S. S. Smith and R. K. Schulze . . . Ed Birge and Floyd Waterfield . . . Burt Hull, Al Horne, and Slats Latimer . . . Miller and David Williams and Oscar Dempsey . . . Paul Kayser and Ray Fish . . . he knew them all. Steve Bechtel called him "Mr. Reed."

These were the men whose work shaped modern pipeline design, who conceived and organized the great oil and gas projects, and who perfected and put to work the techniques which made long-distance pipelining possible.

As assistant pipeline editor 25 years ago, I never ceased to be amazed at the trust these great men placed in my boss. In his presence and mine they discussed new ventures which today are household words but then were still secret, confident Paul would never tip their hand until they gave the signal.

I dined with him at Antoine's as he got two estranged pipeliners to bury the hatchet, ending a feud of a generation. No one else could have done that.

When Paul died last week at 88, he joined his beloved Bonnie who preceded him in death by scarcely a month.

In pipelining as in all of life, Paul Reed walked among the giants. He was one of them. Paul didn't realize that, but the rest of us did.

A hyperseminar

John Jennrich **October 27, 1980**

Last week, Sen. J. Bennett Johnston (D-La.) talked about America's energy needs in 2000 and the regulatory hindrances that stand in the way of achieving those needs.

His conclusion: "You can't get there from here."

I know why he said that.

It's because he was standing in Alliance, Ohio, at the time, along with a bunch of reporters who, along with me, had spent as much time getting there and back as we spent in a classroom listening to seminar speakers.

One thing you can say about Alliance: No matter where you're reading this, you can't get there from here.

The seminar on alternative fuels was sponsored by J. Ray McDermott & Co. and held at the Alliance research facilities of its Babcock & Wilcox subsidiary. B&W has a bang-up research center and I know the staff meant well, but there are some lessons to be learned.

Remoteness and time: If a company is going to hold a meeting in the boonies, make sure the time spent is worth the trip there. First day we had a 3-hour tour that numbed the mind with a blur of 10-minute stops to learn about such subjects as a laser doppler velocimeter and an infrared thermographic imaging system. Some were fascinating, but how much can you learn in 10 minutes? Day 2 brought 16 lectures, so many that often there was no time for questions.

Organization: One example. After criticizing the federal government for single-issue regulation and lack of coordination, B&W handed out seminar notebooks as we left for the bus home. Included was a thick blank pad for taking notes. I gave it to my son; he drew a duck on it.

Spontaneous technical breakthrough: In its only press release, B&W "announced today" the burning of a coal-and-water slurry. Some of us had seen it on the tour. Others didn't because the system failed and jammed the pumps. Announced today? On page 2 of the release reporters found that B&W had already burned 12 tons of the stuff.

Lack of time for journalists to ask questions, a highly orchestrated tour with starched speakers, the notebook gimmick to reduce notetaking and encourage glassy-eyed concentration on top-brass speakers — all these suggest an effort aimed more at impressing the boss than at dealing forthrightly (the only way) with the press.

After all this hyperorganization, maybe it was fitting that B&W sent me off to the airport with a bus driver who, once in Alliance, couldn't get out. After wandering around the beautiful Ohio countryside, complete with horsehead pumps nodding somnolently, and with only 45 min for me to catch a puddle jumper, the driver got the bus stuck in the mud as he tried to back up on a gravel side road.

John's red jacket

Randy Sumpter **December 1, 1980**

There are a number of methods to project how much money an Outer Continental Shelf lease sale will generate for Uncle Sam.

None of them is precise or satisfies a journalist's every need. I ought to know. I've tried them all at least once during the last five Gulf of Mexico lease sales.

The conventional approach is to talk to representatives of the more successful bidding groups and match their comments with the size of the sale acreage, what's known about the area's geology, and what the previous bidding and drilling results have been on adjacent tracts. In addition, I've stooped to eavesdropping on presale shop talk in some of New Orleans' finer hotels and restaurants.

I can't afford to spare any effort. It's my aim to bring the Journal's readers the best presale and sale coverage that I can drum up. That philosophy dovetails perfectly with my boss's favorite slogan, "He who does his job best will get to keep it."

You can imagine then how grateful I was to discover an infallible method of predicting sale results: the color of John Rankin's sport coat.

"I wear red coats when I smell money," says John, manager of the Bureau of Land Management's New Orleans OCS office. In his official capacity, he has conducted all but four of the 41 lease sales held in the gulf since 1954. John has worn out one red suit and three sport coats performing those duties.

The first of the fire-engine red jackets originally was ordered for the employees of a Russellville, Ark., chicken feed store. John, a Russellville boy, couldn't resist when he saw one of the coats in the store window. He bought it, thus beginning a long and unusual tradition in OCS leasing that they probably haven't heard about in Washington, D.C.

John and some of the bidders are willing to admit the red jackets have become a superstition.

He wore the latest jacket in September for Sale A62, which broke a high bonus money record. After counting the receipts, John pledged to retire the coat.

Last month, he showed up at Sale 62 wearing the same jacket. That sale offered only half the acreage that A62 did — and off Texas to boot. But the results easily passed the $1-billion mark.

If I were the new secretary of Interior, I would make sure John had a fresh red jacket before each sale. Then I'd get the measurements for the three other OCS office managers and go shopping in Russellville.

It's never dull!

Robert J. Enright **December 29, 1980**

This issue's Worldwide Report caps a full year of OGJ reporting and writing on world petroleum developments.

In mulling this year's version, we're reminded how difficult it is for even the most clairvoyant of world petroleum experts to see very far ahead.

For instance, who would have dared predict a few years ago that Petroleos Mexicanos would now be producing more than 800,000 b/d from its Gulf of Campeche oil fields — an average 30,000 b/d from 27 Campeche production platforms? And that most likely is just a beginning. Nearly all of that production is from the Cantarell complex of fields. And there are dozens of attractive structures in the immediate area which haven't yet been touched by the drill.

And who a year ago could have foreseen that Iran and Iraq would be at war, with a loss in world crude exports of about 3.5 million b/d? We wouldn't have been too surprised to see Syria attack Israel (aided and abetted by Libya) and perhaps to see some Arab producing countries along the Persian Gulf dragged in. But Iran and Iraq? This sudden eruption in two of the Middle East's largest oil producers was the biggest single event influencing world crude supply/demand this year.

And who a year or so ago would have predicted that U.K. North Sea oil production this year would be a miserly 2% higher than in 1979? The increase last year over 1978 was a resounding 21% plus. Brent, Ninian, and several other fields were expecting major increases in production, and several other fields were expected to produce their first oil this year. What happened? Operators in Brent and other fields have had fits trying to inject associated gas the government won't permit them to flare. Consequently, their oil production has been limited to far less than the volume that had been expected. And several new fields previously expected to be on stream this year didn't make it. Texaco's Tartan and BP's Buchan were among those falling victim to delays.

And who could have predicted in the heady days following the Conservatives' election victory in Canada that they would be promptly booted out and that the triumphant Liberals would just as promptly adopt an absolutely disastrous energy/tax program from the standpoint of the Canadian petroleum industry?

Finally, who could foresee that perhaps the largest field yet in the West African theater would be found off the Ivory Coast? We can't say with assurance that it is or isn't West Africa's biggest offshore field because the Ivory Coast government has clamped down on release by the participating companies of information on this field. But we can say with some assurance that it appears to be a good one.

There's something else we can say with certainty: Covering worldwide oil is a lot of work — but it's never dull!

We get queries

Theodore Barrington **January 5, 1981**

Ninety percent of the queries the Journal receives are legitimate, intelligent, and we're glad to try and answer them. The balance are in another category. Some examples:

A lady called Engineering Editor John Kennedy wanting to know who the president of Saudi Arabia was.

"They don't have a president, ma'm," Kennedy said. "They have a king."

"Well I demand to know where I can write him," the lady said. "I'm going to give him hell about oil prices." Given his address, the distressed lady vowed she was writing the king that afternoon. (Two weeks later the price of oil went up.)

Then there are those wide-eyed innocents who are about to leap, virtually unborn, into the oil-trading arena. They call and whisper conspiratorially to Economics Editor Bob Beck: "Say, what's a good price for Egyptian crude? I've got a chance to buy a tanker-full." Beck has fielded several of these and usually passes them on to the crude-supply division of an oil company.

Jim West, our news editor, takes the intellectual calls. A man rang him up wanting to know what the differences between an operator and a refiner was. Of course, he should be forgiven — he was calling from the Energy Department. And then there was that reporter from a major metro daily who asked, regarding rumors of a multimillion-barrel oil shortage in the 1990s: "Will there be a shortage of barrels, too?"

My favorite calls are those that pine for riches. These go to Exploration Editor Jack McCaslin. The other day a lady called him and said, "They're drilling a well across the highway from my property. Does that mean the oil stops at the highway?"

"Not necessarily," answers Jack.

"Then how can I get them to drill on my land?"

"If they find oil across the highway they'll be right over," Jack replies.

And so they will.

Perhaps the most charming of recent calls was one directed to Pipeline Editor Earl Seaton. A lady wanted to know: "How many miles of downhole pipeline have been laid in the Gulf of Mexico?" Seaton kindly disabused her of that term, and also translated the malapropism of horizontal casing.

Some years ago, Seaton received our most curious reader response. He got a Christmas card from a man he'd never heard of, no return address, addressed to Early Season. It was inscribed, "Merry Christmas and Happy New Year. But I can't speak for my wife."

Readerland is varied.

Our new look

Max L. Batchelder **January 12, 1981**

It is somehow appropriate that the words you are reading right now make up the first assignment anyone has written for the Journal using this battleship-grey, green-eye-blinking, type-producing cyclops before which I am seated. It is the keyboard and viewing screen of a computer typesetting device and I have this terrible feeling it is smarter than I am, but is too polite to say anything about it.

Until lately, the copy for OGJ was mostly produced by editors sitting in front of assorted typewriters, staring at blank paper, wondering how to get it all into the first sentence. Once the words got on paper, a great mechanical type-setting monster clanked out rows of hot, heavy metal slugs used to print those words — just the way it was being done decades ago. Our presses have long since been upgraded to the best offset equipment in the industry, capable of printing the highest-quality material brought to it. But tapping out the stuff on a typewriter and setting it into metal type — that part stayed the same.

But now those old gargantuan metal typesetting machines are gone. The typewriters are out. The long galleys that used to smear fingers with printers' ink no longer plague us. No more hot metal. Not a clank left in the house. Instead, you now hear the whisper of soft clicks as fingers quest across an electronic keyboard, feeding words into a computer that produces type photographically on paper, with no hot metal, no slugs of type, no fuss, no muss, and with practically no noise at all. The editor suddenly has his fate in his own hands, because now, for all practical purposes, he makes his own type.

The possibilities opened up by the new procedure are tremendous — you will already have noticed all kinds of differences in the issue of the Journal you are holding. To match our new technology, we have come up with a totally new format with the help of magazine design wizard Jan White, slicking up most everything in the book — headlines, tables, illustrations, captions, the cover, even the face of the type you are reading at this moment. The new system will also let us do far more things with formulas, tabular material, and equations within the regular text than ever before.

There is, of course, a purpose in all this modernization: greater efficiency for us, a more readable and useful product for you.

It means doing things a new way, letting go of old habits. But that's no trouble for Journal editors — after all, we've been dealing for most of a century with an industry that constantly develops new technology, employs new operational methods, and uses new, more sophisticated equipment.

So here we go with a new look, a new procedure, a new series of deadlines. For nostalgia, I may still yearn for the sound of a clanking Linotype machine, but already I can tell you I wouldn't trade this slick new keyboard for all the clattering typewriters in the world.

A memorable assignment

Ted Wett **April 6, 1981**

It's not unusual for an Oil & Gas Journal editor to travel in search of the material that makes up one of OGJ's special reports such as the petrochemical number in this issue. In fact, travel is expected.

But this year will go down in my diary as fitting the unusual category — and then some. Not so much for the geographical area covered, although it was extensive — from coast to coast in the U.S., to Europe, and to the Persian Gulf. More so, it was the variety of sights and situations that make it memorable.

First, there were the contrasts in sleeping accommodations. They ranged from a delightful ocean front motel in Santa Barbara, Calif., complete with crashing surf; to a hotel in Paris, "in the winter when it drizzles," and where the spongy mattress couldn't be helped much by the old desktop bedboard they dug up to try to ease an aging editor's aching back; to the rock-hard bench in the Bahrain airport where intermittent periods of fitful sleep helped pass the night waiting for the plane's "hydraulic problems" to be corrected. Memorable indeed!

More disconcerting was to read about the gunfire, kidnapping, and other violence that took place not far from the Beirut airport almost simultaneously with a brief stopover there to change planes.

Memorable too — but much more satisfying — was a visit to the first ethylene plant based on feedstock recovered from associated gas to go on stream on the Persian Gulf, in Umm-Said, Qatar.

As our incoming jet sailed over the rooftops in the center of that small nation's capital, Doha, one felt he could almost reach out and touch the top of the Grand Mosque's clock tower. A short drive from the airport brings you to the sparkling, efficient-looking petrochemical complex rising out of cement grey sand. Cement grey sand? Where are those lush brown dunes that I've learned to associate with the Middle East?

They're around, I found out later. Southwest of Umm-Said, moving toward Saudi Arabia, the sand color shifts from grey to brown. And there they are, dunes beginning as if a sign were posted saying, "Start here," and 20 ft high to boot.

Even Qatar's place in the development of petrochemicals in the Middle East is unusual due to a combination of circumstances. If the Iranian revolution hadn't taken place, the first plant of this kind to go on stream could have been at Bandar Khomeini, when it was still Bandar Shahpour. If the Iran/Iraq conflict hadn't started, it could have been at Basrah. But Qatar it was, and memorable.

So preparation of this year's petrochemical report resulted in some unusual experiences. Who knows? Maybe next year, with the world of petrochemicals taking on a more international profile all the time, will make this year look like just another routine assignment. But I'm not really looking forward to trying to top this one in any way but editorially.

The days of Old Betsy

Randy Sumpter **June 1, 1981**

There was a time when a good cannon and an accurate cannoneer were essential for refinery fire fighting chores.

They were called upon at some refineries when one of the heavy gauge sheet metal and rivet storage tanks used earlier in this century caught fire. As the crude in the top portion of the tank burned, water, which had separated to the bottom, began to boil. That created pressures the old tanks just couldn't handle.

Eventually, something had to give. Vapor pressure from the boiling water would blow up the tank. The results could be catastrophic, depending upon how large and how far away neighboring tanks were. The most practical solution was to somehow poke a hole in the smoldering tank before an explosion occurred.

The safest — but not entirely foolproof — solution was a well placed cannonball.

On one such occasion during the earlier days of Gulf Oil Corp.'s Port Arthur, Tex., refinery, the black powder and cannonball solution created quite a stir. The refinery's fieldpiece — Old Betsy, a relic from the War of 1812 — was trundled out. The cannoneer, working on the theory that if 1 lb of black powder did a good job 4 lb would be better, loaded Old Betsy, aimed at the offending tank, and touched off the fuse.

The cannonball punctured the burning tank all right but continued to sail on for several hundred feet and pierced another storage tank. The refinery crew was able to hold further damage to a minimum.

Old Betsy was fired once more — Nov. 11, 1918 — to celebrate the World War 1 armistice. On that occasion, the blast shattered storefront windows along several blocks of Port Arthur's business district.

Management had had enough. Betsy was plugged and mounted in front of the refinery's administration building where it sits to this day.

Tank design has come a long way since then. Fail joints, welded seams, lighter weight walls, and injectable foam have eliminated the need for cannonballs and cannoneers in the oil patch. And industry now is honing other refining skills and engaged in major retrofitting projects to handle the bigger slice of heavy, high sulfur crude entering the marketplace. Tank design hasn't been forgotten in Oil & Gas Journal. Our coverage in this issue from API's midyear refining meeting includes a paper on the design of floating roof tank drains. There isn't a word in there about black powder.

Regional shock

Patrick Crow **June 15, 1981**

The Russians have a folk saying: "Every rat loves the swamp in which he was born."

I realize now that holds true for me. I moved to the Washington area 2 years ago after living all my life in Oklahoma, and I've had some adjustment problems that Russian rats never dreamed of.

A typical westerner, I feared and found the East to be claustrophobic. What bothers me is not the urban congestion (how could anything be worse than Houston?) but the lack of wide open spaces and wind in my face.

Tall hardwood forests block the vistas and, I suspect, the wind. The seashore is my therapy. It's always windy, and the horizon is unobstructed. Even the sand in my shoes reminds me of Oklahoma. I'm usually the only guy on the beach staring inland at the sand dunes, not the ocean.

Back in the Southwest I could find my way anywhere with a map and the knowledge that highways and streets generally were laid out in north-south grids.

However, tradition — not logic — reigns in the East. The traffic arteries here follow paths Indians or cows found most convenient 200 years ago.

Street names also can be illogical and inconsistent. Near my home, for instance, Ox Road becomes Chain Bridge Road, also known as Route 123, after crossing Little River Turnpike. Ox Road is for cars, Chain Bridge Road has long been concrete, and the "turnpike" isn't. These things don't help when you were lost to begin with.

I'm also suffering from regional food shock. In Oklahoma the mainstays of my diet were hickory smoked barbecue, chicken fried steak, and sour cream chicken enchiladas surrounded by refried beans and rice.

Here, whenever I order the local barbecue I seem to get yesterday's dripped beef. The sauces have all the finesse of peppered ketchup. As I eat this mush, I gloomily suspect that Texans have cornered the market on hickory.

Chicken fried steak or true Tex Mex just can't be found.

The regional specialty is Chesapeake Bay blue crab. You boil them, attack the shells with a hammer, and then laboriously salvage a thimblefull or two of meat. You're warned to avoid eating the innards called "the window's fingers" or anything yellow. Yuck!

It takes about a bushel basket full of these things to fill you up, because you get pretty hungry in the 1 1/2 hr it takes you to pick out your dinner.

If that's too slow, restaurants abound — particularly ethnic restaurants. In Washington, for instance, there are six places to eat genuine Ethiopian food. I intend to wait, however, until the next time I'm starving in Ethiopia.

You can take the boy out of Oklahoma — but not his stomach.

An unfair advantage

Gene T. Kinney **September 14, 1981**

When you a find a presidential candidate with the courage of your convictions, a word of caution: There's always the risk that he may be elected and carry them out.

The oil and gas industry is facing the ultimate test of its principles. A politician who takes them seriously now sits in the White House.

When J. Cordell Moore was assistant secretary of Interior in the Johnson administration, he discovered something. There was a certain pattern in the petitions lodged with his office.

"No one," he used to say solemnly, "ever asked me for anything more than an unfair advantage."

His counterparts in President Reagan's administration have made a similar discovery. That attitude crops up in one form or another on all major issues. Whether it's decontrol of oil or natural gas, acceleration of offshore leasing, or windfall profits taxes, the story is much the same.

Decontrol of oil prices is good for producers and for national refiners, who have been sheltered from the world market by price controls, subsidies, or both, and now face competition from foreign plants with lower costs of taxes and pollution control.

Gas decontrol, too, is popular in principle. But it scares pipelines and distributors, who fear destruction of their markets and loss of high-priced Alaskan gas unless it can be averaged in with regulated supplies. And decontrol of all gas gets a lukewarm reaction from deep-gas producers, for whom deregulation is much more profitable while other types remain regulated.

Next to decontrol, industry has called loudest for more acreage to drill. But now that the administration proposes to open up the entire Outer Continental Shelf, some companies complain that's too much.

Everyone, it seems, wants the government off their backs. But few want to be left entirely to the mercy of the marketplace.

Political sophisticate that he is, President Reagan surely understands this. But he must wonder at times if it isn't rampant free enterprise that some businessmen fear most.

No candidate in modern times has been elected on a platform so closely resembling the agenda of business. But this President has discovered, like others before him, that the petroleum industry is not monolithic. There is no single industry view. The industry is made up of many divisions, each with its own segments and even factions within those segments. What's good for producers and consumers may in the short run drive refiners to the wall. And what's good for some producers and refiners may be bad for others. This is one of the great frustrations of politicians who share industry's free market philosophy but need a clear signal from industry in order to act.

For oilmen, Ronald Reagan represents a rare opportunity — a president with the courage of their expressed convictions. As he weighs future legislation on natural gas and taxes, he needs expert help in framing policy to serve the national interest, not lobbying to aid one narrow group or another. He can do without the kind of special pleading that Cordell Moore recalls — more in sorrow than in jest.

The hazards of beards

Roger Vielvoye **September 21, 1981**

Oil & Gas Journal has few editors with a full set of face whiskers. Only three spot a beard.

Growing a beard has never held any attraction for me. A couple of flirtations with droopy mustaches did not eliminate the daily chore of shaving, and hair on the upper lip rapidly turns into a filter for the more succulent items on the menu.

Until recently it never occurred to me that the state of a journalist's facial hair — provided, of course, that it was neat and well trimmed — was of any consequence.

But during a visit to West Germany's Wintershall AG with a number of other European journalists the benefits of going clean shaven became very apparent.

The first part of the itinerary went without incident — a tour around Wintershall's Erdolwerke and a look at work on secondary recovery in the reservoir engineering laboratories. On then to visit gathering and treatment facilities for small oil fields that are dotted around the Emsland area of Northwest Germany.

Then it happened. As the party was about to board the bus for Wintershall's gas purification plant at Duste, one of our hosts broke the bad news: The two members of the party with beards could not enter the plant.

Silence. Smiles and sniggers from the clean shaven majority. Stunned amazement from the hairier members of the group. What had they done? Why the sudden discrimination? They momentarily had that persecuted air of a wronged minority.

All was then revealed. West German safety regulations require everyone entering the plant to carry a small respirator because toxic fumes might be discharged during an emergency.

Beards, explained our guide, would prevent the respirators from making a proper seal across the face — possibly with terminal consequences.

Grasping the little red breathing devices, the rest of the party began the tour of the pipework and columns that clean up Wintershall's high sulfur gas. Was this the chance to steal a lead on the bearded pariahs last seen gazing at the plant across the barrier at the main gate?

It was not to be. The story, in all its technical glory, had already been covered extensively by OGJ. Back to the bus, where bearded faces waited anxiously for our return. What had we seen? The same pipes and towers that could be seen through the fence, only at much closer quarters.

Next stop was lunch. Would the bearded outcasts be allowed to eat with the more socially acceptable smooth faces? No problem, said the guide, and they were welcomed back into the bosom of the party.

The future revisited

Gene T. Kinney **November 9, 1981**

As the American Petroleum Institute convenes its annual meeting in Chicago this week, oilmen confront a vastly different world from the one that existed 10 years ago.

Consider international crude oil prices. Who could have imagined such a staggering increase during the decade? And who could have foreseen the nosedive in demand? The National Petroleum Council, enlisting the best minds in industry, overestimated 1980 oil consumption in the non-Communist world by almost 25 million b/d.

To get a fresh appraisal of the future for this special API issue, we went to 10 of the top executives and economists in the industry. Fortunately, in spite of the humbling experience of the past decade, they haven't sworn off forecasting.

We believe their vision of the 1980s makes fascinating reading.

Texaco's John K. McKinley dramatizes the enormous impact of past conservation: "It is equivalent to two and a half times the recent daily production of Saudi Arabia." Consultant John H. Lichtblau assesses the future effect: "World oil demand will start recovering next year but will climb at a very slow rate and is unlikely to exceed the 1979 level of about 52 million b/d in this decade."

Several of the chief executives see an end to shortages, barring some new political disruption. As a result of the slackening of demand and the rise in non-OPEC supply, Mobil's Rawleigh Warner Jr. believes the full OPEC capacity is not likely to be called on in the 1980s.

Standard of California's George M. Keller goes beyond that: "Socal is looking for a demand-constrained world market for oil through the end of the century, resulting in a continuation of today's surplus in production capacity." In the view of Shell's J. F. Bookout, the U.S. demand for oil peaked in 1978 at 17.8 million b/d and will decline to 15 million b/d by 1990.

Gulf's Jerry McAfee stresses "the exciting developments in the American oil patch (which) carry a message of fundamental policy importance for this nation: 'No energy problem is so large that the market cannot solve it with a minimum of disruption'."

The opportunities for the U.S. industry are emphasized by Standard of Indiana's T. B. Redmond. He predicts "90,000 wells/year by 1985 and more than 100,000 wells/year throughout the decade of the 1990s." Union of California's Fred Hartley, buoyed by the improved U.S. political climate, feels "the progress we have made in the past 2 years shows that increased energy security lies within our grasp."

Phillips's William C. Douce sees uncertainties, but predicts "the oil industry will react to longer-term incentives and persevere in its efforts to produce more oil and gas — both at home and abroad."

Exxon's C. C. Garvin, calling today's glut transitory, believes the long-range energy problem remains basically the same. And he concludes — as we do — on this cautious note: "It would be dangerous to allow today's euphoria to distract us from what needs to be done."

Keep it simple II

Jim West **December 7, 1981**

Several years ago in this space I wrote a column with the headline, "Keep it simple."

In it I urged writers of news releases and things such as quarterly and annual reports to keep their writing simple and direct by avoiding "in" phrases like "at this point in time" and "adversely impact."

I've seen those phrases in earnings reports that say, "At this point in time it appears that the oversupply situation in petroleum products will adversely impact our third quarter earnings." Translated, that means, "The products surplus likely will trim our earnings."

To no one's surprise, my urgings went unheeded. Writers of all kinds still write as if they were trying to impress instead of inform.

To prove I'm making no progress, one of the latest earnings reports that came across my desk said the company's income will be "negatively impacted" by depressed prices. That means, "We're not going to have lower profits, we're going to show a loss." I think.

I am undaunted by the lack of acceptance of advice in that earlier column. In fact, I'm ready to launch another campaign against a whole new set of silly, useless words that have cropped up in press releases, reports, and speeches.

It's hard to choose the worst words in today's vocabulary, but "prioritize" has to be near the top of the list. That's a horrible example of a perfectly good noun turned into a flawed verb.

You perhaps have seen it used like this: "Shareholders can be assured that management will prioritize its actions in order to achieve corporate goals and objectives." Someone failed to suggest that shareholders would more clearly understand—and thus have their fears allayed by—"We will take our actions in the order of their importance to make sure the money you invested in your company is spent wisely."

I have lots of other words on my hate list. Among them are, "upside potential" and "downside risk." They seem to have become favorites in the business world in sentences like, "Management currently is weighing "the upside potential against the downside risk in refining/marketing." Simply put, that means, "We're trying to decide whether we stand a better chance of making money or losing it."

Some of the words I've grown to dislike most are absolutely harmless by themselves. But they irritate me in combination with other equally harmless words.

Consider these combinations: "old tradition of industry leadership," "a new innovation in exploration," "positive benefit of decontrol," "brief summary of third quarter activity," and "real facts in the lawsuit." None of those modifiers is needed. By definition, a tradition is old, an innovation is new, a benefit is positive, a summary is brief, and facts are real.

Here at Oil & Gas Journal it's my job to read about 10 jillion words/day, distill them, and see that they are reported in a manner that gives you the greatest volume of news with the least amount of reading.

We're doing our best to keep our writing simple and to the point. And we fail our job if we let words get in the way of information.

Farewell to a friend

Gene T. Kinney **January 11, 1982**

Wallace Pratt. Among the world's oil finders, the mere mention of his name is enough to evoke feelings of respect, awe, even reverence.

Few men in any field of endeavor have ever enjoyed such a towering reputation. His place in history, and in the heart of the industry, will still be secure long after time has dimmed the memory of those fortunate enough to have known him.

Some of us at Oil & Gas journal were among that lucky group. We knew him as a faithful reader, prolific correspondent, author, and friendly counselor. Because of that long and close relationship, news of his death at age 96 last Christmas Day arrived here with a special jolt (OGJ, Jan. 4, p. 56). It came in typical fashion, in a telephone call from a mutual friend.

Wallace Pratt "retired" in 1945. That was long before most of us embarked on careers that brought us into contact with him and the industry which he did so much to shape. But his contributions to the industry, the science of geology, and the nation did not end there.

As far as we could tell, his retirement meant only that he was no longer vice-president and a director of Standard Oil Co. (N.J.), an organization he joined in 1918 as the first geologist hired by Humble Oil & Refining Co. Although he severed his official connection with the company, he didn't stop working. He began a long and equally illustrious career as a consultant.

And he didn't stop thinking and writing about the mysteries of oil and gas exploration, which never lost their fascination for him. He continued to be an avid reader of the Big Yellow Book. He knew members of the staff either personally or as names on the masthead and telephone friends. During Frank Gardner's long reign as Exploration Editor then International Editor (before his untimely death in 1975), Wallace loved to call and discuss new plays or discoveries at length. He enjoyed a similar relationship with Frank's successor as Exploration Editor, Jack McCaslin.

We were flattered that he considered Oil & Gas Journal to be his personal information service. When his copy failed to reach Tucson on time, he would be on the telephone to find out why. There were also those embarrassing moments when his subscription expired or we put him on an "inactive" list by mistake, Those incidents, mercifully few in number, stirred recollections of his early days before oilmen liked or trusted geologists and allowed them into the decision-making process in buying leases and choosing drillsites. "Does the Journal also hate geologists?" he would tease. "If so, why do you have them on the staff?"

I remember him best from his work for our 75th anniversary issue, Petroleum 2000, in which he discussed "Oil finding — the way it was." The reader of that article, published in August 1977, can still learn a valuable lesson in how to find oil — and how to write about it.

When Wallace Pratt died, industry did indeed lose one of its giants. And we lost a friend.

An untapped resource

Bob Williams **February 8, 1982**

Here in the West, oilmen often take a novel approach to extracting big but tough-to-exploit energy resources.

At a recent luncheon, I met one of these "build a better mousetrap" types—E.E. "Swede" Larsen. The 63-year-old veteran of the industry has hit upon a new way of exploiting a huge but dormant resource, putting that energy to work, and saving oil and gas firms a lot of money.

The resource base is people—retired oil people with experience, that is. Larson formed ROPE Inc. to tackle industry's acute shortage of skilled personnel and at the same time tap the valuable experience of industry retirees.

Many marginal petroleum projects sit on the shelf, says Larsen, because operators don't have the skilled personnel to review or develop them. That's where ROPE's pool of 250 oil industry retirees comes in.

Projects languishing for lack of skilled employees can be handled on short notice under a one-shot ROPE contract. Or a firm can sidestep no-rehire policies and retain retirees as consultants through ROPE, which carries its own insurance. No fuss about fringe benefits or incentives. No need for a department head to waste time trying to help a weevil get past square one.

It's not unusual for a ROPE team to bring more than 100 years of experience to a job.

ROPE's services cover all aspects of petroleum reservoir, production, and drilling engineering through tertiary recovery, workovers, reservoir evaluation, analysis of equity, risk, and acquisition, and other areas. It also offers training programs for field workers to managers, office to onsite.

Larson, retired from Dresser Industries Inc., and ROPE Vice Pres. Phil Cook, an engineer retired from California's state lands division, carefully screen all ROPE résumés. They have a reputation to protect.

But problems can crop up. A rookie engineer embroiled in company politics might get his feathers ruffled when the wisdom of experience suggests a better approach to a project.

"A lot of times, people are just protecting their jobs," Larsen said. "They tell management what it wants to hear. We tell management what it should hear." That attitude must work pretty well. ROPE gets a lot of return customers. And retirees are, happy, staying in touch without the daily grind while getting an income boost.

ROPE, based in Santa Fe Springs, Calif., is expanding beyond the West Coast. A new office is likely soon in Houston, and other potential is being eyed in Denver and Tulsa.

As companies know only too well in these days of severe shortages of skilled personnel, their most valuable resources are people.

To Swede Larsen, sometimes it just makes economic sense to go after the proven reserves first.

Art and Science

John L. Kennedy **March 8, 1982**

I'm not one who believes the good old days were that great.

Even if there were drillers who could tell you the gravity of the oil in a productive zone by resting a hand on the drilling line, there also were probably a few drillers who couldn't tell from the drilling line whether they were even on a rig.

There was a lot of art involved in drilling in the old days, and if it weren't for the artists that were making hole, then the world wouldn't have a lot of the oil and gas it has now. And a lot of the tools available now that will permit the industry to find even more oil and gas got their start a long time ago.

But I suspect that a lot of what passed for art when it succeeded was called guesswork—or something worse—when it failed. There certainly were opportunities for a little science to creep into many phases of the drilling operation.

Good or not, the old days are gone.

Technology continues to strengthen its grip on drilling operations. Now the driller can let a transducer rest its "hand" on the drilling line and tell its story to a microcomputer.

There's still a lot of "feel" involved in drilling a well, but to do the job the industry has before it, it's going to have to use all the technology available.

It's difficult for me to get a handle on industry drilling efficiency. The more I study well and footage statistics, the less confident I am of my ability to interpret them.

For instance, if you divide the footage drilled in a year by the average number of rotary rigs operating in the same year, the result is "feet per rig per year." If I told you that in recent years this figure declined as the number of active rigs increased, you might say, "So what?"

Even if I said that from 1979 to 1981 the number of feet drilled per rig per year dropped 18%, you might say: "Big deal. The number of active rigs increased by more than 80% during that time. Looks to me like the industry did a pretty good job holding the line."

See what I mean? But the point is that with about 4,000 rigs operating and the prospect of a breather in the rate of increase, now is a good time for the industry to start drilling as well as it knows how. Now is a good time to use all the technology available.

Still, we should remember that digital readouts and automatic pipe handling won't do it all.

Setting a downhole tool at 18,000 ft with a string of steel pipe dangling in a space about the size of a residential air conditioning duct takes more than just skill.

It takes an artist!

Decorating for geologists

Jim West **March 1, 1982**

I have long heard the saying, "Inside every silver lining there's a dark cloud."

That's certainly true in the U.S. petroleum industry today. It's especially true in exploration, the busiest segment of the industry.

Without exception, managers I have talked to are enthusiastic over the unleashing of the country's mighty exploration machine. But their enthusiasm usually is clouded by a persistent worry: "Where can I find enough professional personnel to do the job I want to do?"

Industry's scramble for personnel of all types is apparent in the classified advertising pages of Oil & Gas Journal. Just look at them. They are packed with pleas for almost anyone with professional training.

Fact is, they are so packed that last year OGJ ran a record 1,323 pages of classified advertising. True, not all of the ads were for personnel. But the "help wanted" category sure chewed up a lot of space.

The volume of classified advertising is only a broad gauge of industry's scramble for all professional personnel. Even as we speak, I hold in my hand absolute proof that geologists are the No. 1 priority of those charged with hiring and retaining professional staffs.

The proof? A report from Griswold, Heckel & Kelly Associates Inc. (GHK), an office design firm based in New York. I'll admit that sort of firm isn't a likely source to tap for oil news stories. But I'll take my sources wherever I can find them.

GHK acknowledges that top geologists command salaries and overrides that can total $200,000/year. Those just graduating from college receive salary offers of $25,000/year or more for a bachelor's degree and $35,000/year or more for a doctorate.

Those facts doubtless are well known to personnel recruiters. What may not be so well known is the fact that hefty salaries alone may not be enough to attract and keep geologists. On top of those salaries are piled "office perquisites previously reserved for top executives," GHK reports.

"While once a company could afford to provide a geologist with only the bare necessities of a desk, drafting table, and side chair in a 150 sq ft windowless office, this is no longer the case. Office sizes have been increased to about 200 sq ft and moved to the periphery of the building offering the choicest views."

What's more, GHK says, many geologists have been given carte blanche in decorating their offices with things like walnut or teak executive desks as well as full wall credenzas, attractive wall coverings, and plush carpeting. I'm delighted there has been such a turnaround in explorationists' fortunes.

Many of us oldsters can remember times—in the mid-1960s, for example—when geologists who had been laid off considered themselves lucky to find jobs selling shoes. They were victims of sagging levels of activity in their segment of the industry.

No doubt about it, young geologists are among the elite professionals who are calling the shots in today's business climate. Given the value of the oil and gas they can help find, there are only two chances—little and none—they will ever have to learn how to sell shoes.

Let's wait a few more days

John L. Kennedy **April 5, 1982**

I hope you'll forgive me for not bringing this up before, but I was waiting for you to be surprised.

Now I'm beginning to wonder whether the surprise is coming, so maybe

I'd better bring you up to date.

I'm talking about an article that appeared in the January/February 1978 issue of a magazine called "40", a magazine for the chief executive officer.

The title was "Exclusive Special Report: The Energy Crisis is Over." The synopsis in the table of contents said, "An exclusive feature story announcing a new, proven process that will end the energy crisis. The new discovery for tapping waiting American oil reserves is of such magnitude that it will affect the course of the U.S. economy and redefine world politics."

The amazing "results" of use of a polymer formulated by a Texas company which would be used for enhanced oil recovery were described in the several-page article which said, among other things, that the discovery ". . . has even farther-reaching implications than the immediacy of oil enhancement and will undoubtedly be recorded in the ledgers of history as the point in time when mankind reached a new awareness of the mysteries of the universe."

There's obviously another reason I didn't mention this before. By comparison, it makes what you read in Oil & Gas Journal pretty dull stuff. Besides, the article quoted the inventor thus: "With an all-out effort, the United States could become oil independent within three to five years."

I thought it would make a nice surprise, and I didn't want to spoil it.

Now I'm beginning to wonder. True, he said 3-5 years, and it's only been 4. And true, world oil prices are declining—something we wouldn't have believed in 1978.

Its also true that OGJ's popular special report on enhanced oil recovery projects in this issue shows enhanced oil production from chemical techniques (including polymer) is about 4,400 b/d, up from 3,000 b/d in 1978. That ain't many barrels, but it's a big percentage.

However, the U.S. is not quite "oil independent" yet. Also, I'm not convinced the discovery announced in that article 4 years ago will, in the writer's words, ". . . restructure the thinking of the entire scientific, industrial, educational and governmental communities, and deeply affect the international economic posture."

Still, I don't want to be too quick to judge. We've waited 4 years. Let's give it a few more days.

Haapi Tahafya

Bob Tippee **April 19, 1982**

For those of you who wonder why Oil & Gas Journal writers don't spice their prose with the breezy expressions you may find in other publications, meet an imaginary reader named Haapi Tahafya.

Except for the fact that he is a petroleum engineer, Haapi is not a typical OGJ reader. Born in Gabon, he lives and works there now. English is not his native language.

I invented him and consult him subconsciously when I'm writing and editing stories for the news section.

Haapi, if he really existed, would be among the 14% of OGJ subscribers living outside the U.S. Part of that group consists of U.S. expatriates.

I assume Haapi knows his job as well as our U.S. readers know theirs. But I cannot assume that he has anything more than a literal facility with the English language.

Thus, if I want to tell Haapi that some petroleum analyst is confident about prospects of the oil industry, I won't say he's "bullish" about it. Unless Haapi has dabbled in the U.S. stock market, how can I expect him to make any sense out of "bullish"?

If the analyst were less sanguine about industry's prospects, what would Haapi think if I described him as "lukewarm"?

Of course, the English language is riddled with regional coinages that began as clever metaphors and graduated into standard usage through repetition.

They color the language but too often do so at the expense of precision. Metaphors make for good thinking but, except in the hands of masters, bad writing.

Furthermore, most of these expressions don't cross national boundaries. This has created some real sport between us Midwesterners in the Tulsa office, where all editing takes place, and our trenchant international editor, London-based Roger Vielvoye. For example, Roger recently reported someone was in "a cleft stick." That sent us to the dictionary, where we learned the expression meant an inextricable position.

Of course, we upstarts in the U.S. baffle Roger at least as often as he does us. One of our headlines recently said an oil company had "slated" a project. In the U.K., "to slate" means "to slight." To us, "slate" means "schedule," sort of. Anyway, it's easier to fit into headlines than the more precise term.

If we English-speaking writers on opposite sides of the Atlantic can do such a fine job of confusing ourselves, imagine what any conscious attempt to make our writing less literal would do to readers for whom English is a second or third language.

So we try to insure our news pages don't puzzle Haapi and our other international readers with reports that, for example, an issue has been put on the "back burner," or that some situation has reached its "bottom line," or that the market has gotten "soft" or "sloppy."

Yet these expressions have a way of sneaking into all writing, which is why there are editors. In fact, all the phrases ridiculed in these paragraphs came out of copy during one fairly light week of editing.

When it comes to preventing fuzzy, metaphoric expressions, you have to play your cards pretty close to your vest.

Dieselization, polarization

L.R. Aalund **May 31, 1982**

If you are the refining editor of this magazine, you pay attention to what goes on and what's said at the American Petroleum Institute's annual midyear meeting.

If you also own a VW diesel Rabbit and recently paid $300 to have a gunked-up fuel tank removed and cleaned, you pay doubly close attention to what's said about diesel fuel at this meeting.

This year a lot was said about it, as our news section of a couple of weeks ago disclosed and as the lead paper in our API report in this issue indicates. The fact is that the oil and automobile industries are engaged in a rather serious debate about the quality of diesel fuel.

Detroit and Stuttgart, home of the Mercedes, are apparently getting some bad news from diesel drivers and garages. The motor makers therefore want the refining industry to tighten up its specs, monitor quality closely, jump up the cetane number, and use flow improvers for cold weather driving.

A refiner reports that the recent bitter winter was a disaster for diesel car owners in the Northeast U.S. Some could not use their vehicles for days, simply because the fuel wouldn't flow.

It's not unusual for the automobile and refining industries to be at odds over fuel specs. Detroit won the biggest battle of them all in the early 1970s when the country went the unleaded gasoline route.

However, in this current skirmish, oil appears to hold the upper hand.

Diesel demand is supposed to nearly double during this decade. But the driving force, literally, will be heavy duty trucks, railway locomotives, and other heavy engines. These can generally burn lower grade diesel satisfactorily. The people fooling around out there in their diesel passenger cars with the fancy Mercedes and Cadillac emblems are not the market, just part of it.

It's therefore unlikely that any oil company will spring for a television ad during the Super Bowl game to push its premier diesel. Diesel, however, would make beautiful subject matter for an ad writer.

Boosting the quality of diesel to where the automakers think it should be will cost money at the refinery because some prime diesel components are becoming scarce. This will of course push up the cost of diesel at the pump, taking away one of its selling points.

American auto makers want to keep the diesel wave rolling. These engines permit them to make large cars and still meet mandatory fuel economy regulations.

How this situation sorts itself out in the coming months and years will certainly make some interesting refining—and newspaper—copy.

Reporting the crises

Patrick Crow **June 14, 1982**

We're in the information business at Oil & Gas Journal, and we're chagrined to see misinformation published about the oil business.

We saw a lot of that during the 1973-74 Arab embargo and the 1979 Iranian oil cutoff.

During the latter, I found the hassle of trying to buy a few gallons of gasoline in the Virginia suburbs of Washington, D.C., less frustrating than watching the television networks' shallow reporting of the crisis, including incorrect stories about oil tankers lying off the U.S. and waiting for prices to rise before they landed their cargos.

So I found it very satisfying to read two recent reports by the Media Institute, Washington. The scholarly studies — laden with data, not rhetoric — gently tore TV networks to shreds for poor reporting.

The nonprofit institute reviewed and evaluated each of the 1,462 stories (totaling 39 1/2 hr) which the networks broadcast during both crises. It concluded:

- TV inadequately explained the economic dimensions of the crises. Supply and price of oil were depicted as twin problems rather than as a problem and its solution.
- TV reporters mostly sought government officials rather than outside specialists to explain the crisis. As a result, government was blamed only about 18-19% of the time.
- When the networks discussed solutions to the crises, they used government sources 77% of the time. They interviewed oil industry officials only 9% of the time.
- When the causes of the crises were discussed, fully 25% of the stories blamed the industry for perpetrating a hoax, profiteering, withholding supplies, or other devious acts.
- Among the networks, ABC gave the most attention to the government as a possible cause of the crises and to the regulation vs. deregulation options.
- CBS blamed the oil industry as a cause of the crises 42% of the time,

compared with 34% for ABC and 30% for NBC. CBS also was the most inclined to call for more government involvement.

• NBC relied the most on the government as a source of information in both crises. Its coverage had the most political and least economic focus.

The networks have ignored the studies, which received too little attention elsewhere, too.

To generate discussion, the Media Institute may distribute its findings to 10,000 libraries and 50,000 opinion leaders nationwide.

Thanks to the Media Institute, another oil crisis is likely to be better reported on TV.

And you can count on Oil & Gas Journal to maintain its reputation as the best single source of information on the industry.

When the media calls on our readers, their script will be in the Big Yellow Book.

More than a little progress

Ted Wett **June 28, 1982**

A billion here, a billion there, and before you know it you're talking about some real money.

In the face of temporarily abundant supplies of conventional fossil fuels and soaring construction cost estimates, many prospective private sector participants in synfuels projects have backed off. Companies have been forced to defend their balance sheets and abandon the luxury of serving a public interest they once regarded as an attractive byproduct of synfuels ventures.

For many who recognize that there will be a future need for these alternate energy sources, enthusiasm still exists. Although their ranks have thinned dramatically, there are still marchers to the distant beat of synfuels drums.

Few doubt that the energy potential offered by synfuels will eventually be needed and profitable. The problem is, according to Royal Dutch/Shell's senior managing director, D. de Bruyne, "how to get from here to there without losing your shirt in the process." Royal Dutch/Shell recently scrapped plans for coal gasification plants in Holland and West Germany.

Standard Oil Co. of California is still pursuing its goal of eventually producing 100,000 b/d of shale oil but has adjusted its schedule to place early emphasis on research and development. Socal Chairman George M. Keller, says, "We're not going to let what we view as a short term market situation greatly delay development of oil shale and other alternate energy resources that are needed for the future."

But economics rears. When the Cathedral Bluffs project (Occidental and Tenneco) was reevaluated in late 1981, it was decided that, "based on increased construction costs, declining oil prices, and high interest rates, the project as contemplated did not meet appropriate economic criteria."

Once viewed as growing to maturity to form the ideal way to guard against continuing increases in OPEC oil prices, the synfuels industry now appears virtually stillborn. At the very least, its development has been seriously stunted.

We said 2 years ago that the status of the synthetic fuels industry gave us a feeling of deja vu. We had been through a similar burgeoning of synfuels projects in the early 1970s, with most of them fading away once oil supplies again became plentiful.

But not everything is the same. This time there are commercial projects that are nearly complete, and some that are still under serious consideration. We've made a little progress.

Also, as we learned through the contacts made while gathering materials for this week's synfuels report, a solid base of technology, experience, and enthusiasm has been formed on which a synfuels industry can be built.

Maybe we've made more than a little progress.

Rainbows

L.R. Aalund **October 4, 1982**

Is that a rainbow on the front cover of this magazine? Well . . . yes it is. And it and other aspects of this picture cost a number of people in Denver, Tulsa, and Houston some busy hours.

We chose this picture not because of the rainbow but because it shows a new resid cracker that Total Petroleum has built at its Ardmore, Okla., refinery. This is the subject of an article in this issue by several people that have been deeply involved with this development.

The first discussion about this article took place 2 years ago this week when I went to Alma, Mich., to propose an article to Bob Dean, senior vice-president of Total Petroleum, on the company's resid cracking technology. But the article got into final form only after I flew to Denver, now Total's U.S. headquarters, for 1 day in August.

This is when I first saw the cover picture. My first irrational reaction was that the picture had been staged. But you obviously can't stage a 40,000 b/d cat cracker.

My second reaction was that the picture was too full of symbolism: "pot of gold, etc," or emotionalism, "look for the silver lining." But the Total people assured me that the rainbow just happened to be there when the photographer took the picture.

I reasoned that rainbows are a fact of life that you have to live with now and then, and that this was an honest, illustrative picture.

But once the picture got to Tulsa in early September, Bob Enright, our executive editor and cover "czar" had a more fundamental objection — our nice picture had too much of a vertical format, while our cover is horizontal. The necessary cropping would ruin the picture, Enright thought.

I agreed, even though the unit's builders in Denver thought it didn't suffer any great harm.

Nevertheless, I then decided to run it within the article without cropping and get another picture for the cover — until I was told I could not get color on the pages dedicated to the article. I didn't at all like the idea of running a rainbow in brilliant black and white.

I therefore in desperation took a look at the picture in a mocked up cover and decided it looked fine. Enright agreed but then wanted to know if it also illustrated the refining index, which is another feature of this issue. I assured him that resid cracking was a growing part of the index.

He then wanted to know why, if the pictured unit was new, it was so rusty.

I put the question to Bob Dean, and he said, "Because we didn't paint it, and because we like earth-tone colors."

Oil crises revisited

Patrick Crow **November 1, 1982**

I've used this space in the past to headline the Media Institute's evaluation of how well U.S. television networks reported the 1973-74 Arab oil embargo and the 1979 Iranian oil cutoff.

In a nutshell, not very well.

The nonprofit Washington study group viewed every story each network aired on the energy shortage subject during those periods. It offered no conclusions, only hard data which showed how superficial and unbalanced the networks' reporting was.

I can forgive the networks for the limitations of their medium and for occasional failures, but their arrogance irks me. Their reaction to this serious indictment of their reportage has been to ignore it completely.

When Media Institute published its first reports, I asked each of the networks for a reaction. I've been a reporter for 16 years and have covered just about every kind of story, but I've never been given such a thorough runaround.

Daily for 2 weeks I sought statements from all three networks at their offices in Washington and New York. I was referred here, there, and everywhere. No one would comment. No one would even say, "No comment."

The same networks which demand full disclosure from government and oil companies "because the public has a right to know" wouldn't admit they might have failed in that same obligation to the public.

Unfortunately, the Media Institute's studies also received a "ho hum" reaction from the wire services and major newspapers. Only TV Guide covered them in any depth.

The third and final volume of the study was released last month. Unlike the others, it contained opinion. It was pithy and biting. But it was ignored, too.

In it, Thomas Hazlett, an economist at California State University at Fullerton, wrote, "Network television news coverage of the oil crises failed to grasp the story being covered and to challenge those in the public and private sectors who manipulated energy policy to the detriment of the common weal. The uncontested fact is that crucial public policies did not accomplish their goals because they ignored the consequences of market forces, and television news failed as well because it ignored this critical aspect of the story."

Granted, the oil crises are history now. U.S. crude prices have been decontrolled. Government allocation rules have expired. So why should the oil industry care?

Media Institute Pres. Leonard Theberge replies: "We are seeing history repeat itself in news coverage of the natural gas deregulation issue."

Recognizing lawmakers

Patrick Crow **December 6, 1982**

A friend and I often ride the same bus in the morning and share our opinions on life and politics as we walk up Capitol Hill in Washington.

"Have you ever noticed how senators don't walk like ordinary people?" my friend, an aide to a senator, asked recently.

"They walk very erect, almost to the point of leaning over backwards. And they tend to walk slower than other people. Perhaps it's the posture."

I knew in an instant he was right. They do walk like they've just been to a chiropractor.

As we compared notes, we agreed on two other ways to spot a U.S. senator in a crowd.

The senator always wears a dark suit with a white shirt and red tie. Always. Senators who wear sportcoats aren't running for reelection.

The senator is always a half-step ahead of his entourage as they walk down a hall. Anyone walking abreast of the senator is either an aide who has seriously overestimated his value to The Man or a constituent who has let the senator's words of praise go to his head.

There are some exceptions: "little s" senators around who want to remain commonplace people. Former Oklahoma Sen. Henry Bellmon was one. Of course, "capital S" Senators are made, not born.

"Have you also noticed that Senator So-and-So has learned how to stroll senatorially?" my friend asked.

The individual (not my friend's boss) is a brash freshman who bobbed up and down when he first walked the halls of Congress. Folks mused that the man had worn cowboy boots all his life and was having trouble adjusting to lighter dress shoes.

Always a good imitator, So-and-So now walks senatorially. He's not any taller these days. He just seems so.

The House of Representatives is an entirely different world. There are 435 congressmen there, so they have to work a little harder to get noticed. Thus sportcoats, flashy ties, white shoes, and cowboy boots are permissible.

The only House members who wear dark blue suits and red ties are the ones aspiring to the Senate.

Unlike the sedate Senate, the atmosphere in the House is like the Elks Club on Saturday night.

House members tell jokes, laugh loudly, and do a lot of backslapping as they ride the underground trams to the Capitol Building or gossip in the halls. They all seem to know one other, even if no one else does.

To help, the leadership gives the freshmen fancy, oversized lapel pins so they won't be kicked off the "members only" elevators and the police will stop traffic on Independence Avenue so they can cross. When they stop wearing the pins, it means they think they have arrived.

These observations may seem trivial, but with so many staffers, lobbyists, and reporters on Capitol Hill it's important that congressmen be recognizable as such. As you can tell, there's little problem.

The more, the merrier

Robert J. Enright **December 13, 1982**

We are read not just in petroleum headquarters in Beijing but in the oil fields, refineries, shipyards, petrochem plants, petroleum institutes, and research facilities throughout this vast land. We documented this in touring Chinese oil installations in the field and in visits at our booth last March with many of the 160,000 from all segments of the industry and the far corners of China attending the big oil show in Beijing sponsored by the American and Chinese SPEs.

160,000?! Might that be all the oilmen in China (that, after all, is many more than make up the entire industry in some countries producing a good deal more oil)? Not on your life. If there's one thing China has lots of, it's people. And oil obviously has its fair share.

Take Renqiu oil field. This is China's second largest single field (ranking only behind Daqing) and its newest major field, found in 1976 on the high plains 105 miles south of Beijing in Hebei Province. The field takes in some 12 sq miles and produces about 240,000 b/d from pre-Cambrian dolomite pay up to 870 m (2,854 ft) thick.

The production operation is conducted by the North Branch of China Oil Operations & Development Corp. and is largely self-contained. It has its own drilling rigs (19 of them), its own laboratories, its own living quarters, its own outdoor produce markets. A city has sprung up around the headquarters to house the oil field workers — all 70,000 of them! That's the figure given us by North Branch officials during our tour of Renqiu. Of that number, we were told, 4,000 are "technicians" and 1,500 are engineers and geologists.

Seeing China as a major new source of crude and markets for equipment and services, the international petroleum industry has responded with great curiosity about the Chinese industry. In this issue's special China Report and throughout the year, OGJ is doing its best to keep oilmen up to date on China and the rest of the world, too.

Whew, it's done!

Roger Vielvoye **December 27, 1982**

Every year the job of compiling the Worldwide Report gets tougher.

And it's sad to have to report that generally as state companies become more and more involved in collating the statistics on production, the efficiency with which they are returned to OGJ headquarters in Tulsa diminishes.

Of course, there are many state companies that conscientiously fill in our survey forms and send them back before the deadline, just as there are private companies that are equally forgetful or late.

Most operators, however, realize that publishing such an extensive document can be achieved only with the full cooperation of the industry. And once again we thank everyone involved for their efforts.

It's not always forgetfulness that causes confusion for OGJ compilers of this report. What insurance adjustors call acts of God often come between Tulsa and the figures.

Last year a sabotage attack on the Petrangol refinery in Angola diverted the attention of local personnel from the job of dispatching the statistics.

This year a fire which destroyed much of the complex used for production records storage at Petroleos Mexicanos headquarters in Mexico City caused some delay in the Mexican figures — and nailbiting by our Worldwide Report team.

Argentine statistics failed to arrive, despite telex and cable pleas at deadline — perhaps due to ill feelings over the Falklands fracas. We nearly suffered the same fate with Brazil, thanks to a mailboy in one of the Petrobras U.S. offices who found it easier to dispose of mail in the wastebasket than to deliver it. (He's no longer around.)

And changes in sources for country by country data occur constantly and cause no end of trouble. Take Ecuador. City Exploration of Ecuador, a faithful reporter on its production in that country in the past, simply disappeared. It took a good bit of detective work to determine that this company not only had moved to different offices in Houston, but had changed its name (now Central Universal) and sold its Ecuadorian production to Clyde Petroleum, a British-owned firm.

Iran and Iraq have regularly reported their total production by fields, but now don't. We can understand that.

Countries involved in shooting wars, however, aren't the only ones reluctant to part with detailed figures. With OPEC production ceilings springing up for the first time during the first 6 months this year and some quota busting going on, some other producers required much prompting.

The Worldwide Report is not made up of production figures only. It also gives the most detailed breakout of refining capacity. And in a year when processing units have been shutting down and changing owners with disturbing regularity, this part of the survey takes on added importance.

Director of Editorial Surveys Ailleen Cantrell had to send out more reminders than ever before to refinery operators but finally got almost a 100% return.

Financial advice

John L. Kennedy **February 21, 1983**

Our Drilling Technology Report in this issue features, not surprisingly, some important drilling technology. It also features the results of one firm's study of the outlook for crude prices and rig activity.

We feel this article is appropriate, considering the industry consensus that technology is much more fun if you can make a profit using it.

In addition to this special report, I want to pass along a few ideas that have much broader application than just the world of day rates and active rigs. Use of these ideas need not be confined to any one business; rather they provide a philosophical base upon which one can confidently lay the necessary details of a specific plan. With slight modifications for individual situations they can be useful in today's business situation.

For instance, business troubles seem to get more attention during recessions. But it is important to remember that not everyone has difficulties when business is down. Keeping this in mind and putting the best possible face on apparent problems can be the first step in laying a foundation for recovery. For instance, I've said to myself in the past:

"I'm not worried about a depression. I went bust during the boom."

With the proper attitude, a plan will begin to gel. Consider, for example, that slow periods are also a time to plan for the future, a time to look for opportunities. Opportunities abound in times of recession, especially if we call problems "opportunities" as we are often instructed to do. Resist the temptation to respond negatively to this idea. Don't say:

"I don't need any more opportunities. What I need is some 'bidness'!"

Much more positive thinking is called for. Remember that the smart guys buy stocks or real estate when prices are low. Why not invest in the oil business when prices are low? It just makes good sense, especially if you have any money or credit.

One of my college-age daughters puts it this way: "When the going gets tough, the tough go shopping."

Words to live by, indeed! This aggressive approach will keep you from being left behind when the next boom starts. As stock market investors say when the market starts to run up: "Happiness is not having to say you're short."

Believe me, these guidelines can help. I've used them and I feel I have the future pretty well mapped out.

This kind of planning — and just one big score — will help you, too.

Head Start

L.R. Aalund **March 21, 1983**

It's rare to pursue an exclusive news story and not have to worry about the competition beating you to it.

That was the case with the refining news in this issue that is based on the interpretation of data gathered for our annual survey of Canadian and U.S. refining capacity.

Until this issue appeared today in most places, few people knew the exact magnitude of the staggering drop in U.S. refining capacity in 1982 and where it happened.

We found, among other things, that capacity fell by 1.5 million b/d and that more than 40 refineries shut their gates. That means there has been a drop of 2.3 million b/d in U.S. capacity during 1981 and 1982.

I don't want to sound smug about this news story. There probably were some individuals or organizations that in February had figured out part of the capacity scene. But they didn't share it with us. Unless you print it or tell someone, it isn't news. It is intelligence or perhaps a "database," the ubiquitous new buzz word for 1983.

Also, there are some people lying right now on the beaches of Cozumel or Acapulco that would give a big yawn if you flashed the U.S. capacity drop figure before them.

But for today, at least, it is real news in the circles we move in.

And we don't necessarily have to rely on our own experience to make this judgment. We get help from some of our callers. This year a larger number than usual started calling me or Ailleen Cantrell, Oil & Gas Journal's director of editorial surveys, 2-3 weeks before publication, hoping to get an advance on the survey numbers.

Most realize they are asking for just a bit much. Therefore, opening ploys can be ingenious. One lady with a charming voice began with, "Are you all doing that refinery survey you did last March again? If so, we would like to get a few numbers." I wanted to say I vaguely remembered it and the other 10 or so I have struggled with. But I didn't. And in spite of the lady's charm, I didn't give away the crown jewels.

Then there is the head-on-type guy in a hurry who simply asks for the totals. It's easier to say "no" to him.

It is obviously not in our best interest to dribble out our products — news, information, and statistics — weeks before publication.

Furthermore, I believe our subscribers — all of them at the same time — should get the first shot at this information if they want it.

The Workhorse

John L. Kennedy **April 18, 1983**

The fact that injecting water into oil reservoirs might increase oil production reportedly was discovered accidentally over 100 years ago. It has also been reported that introducing water into an oil reservoir was illegal in much of the U.S. until about 1920.

The Journal's special report in this issue indicates how much progress has been made since then. The industry now understands all the complexities of how several fluids flow simultaneously in a heterogeneous medium, exactly how much water should be injected in which type reservoir, and precisely what the results will be.

Well. . . maybe not all. Maybe not exactly. And sometimes not precisely.

But it's only been about 100 years.

It's relatively easy these days to build a sophisticated computer model of a reservoir to be waterflooded which details every possible reaction to hypothetical changes in physical conditions, crude prices, and politics. Knowing whether the injected water will go north or south and whether the oil will go with it, however, still poses some problems.

Even after 100 years.

But look at what waterflooding has done for oil production! I'd like to be able to say how much oil has been produced in the last 60 years because of waterflooding that would not otherwise have been produced, but I have no idea. I have heard it said that more than half the oil produced in the U.S. comes from reservoirs that are waterflooded. But I can't vouch for that.

I think you'll agree, however, that waterflooding can be credited with recovering vast amounts of oil.

Now, if we could just recover the other two-thirds of what has been found.

The industry has several ideas about how to get some of this oil. But it will be awhile before any of these can hold a candle to what waterflooding has done. The Journal's 1982 survey of enhanced recovery projects, for instance, estimated that enhanced production from micellar-polymer, polymer, caustic, and other chemical EOR processes totaled about 4,400 b/d. Enhanced oil recovery from all methods was estimated at about 375,000 b/d; almost 80% of that was produced from steam drives.

That's after a few decades of research and development and a 10-fold increase in the price of crude.

It's obvious that making a dent in the 300 billion bbl of remaining reserves is going to be excruciatingly slow. Maybe it won't take 60 years; on the other hand, we've been at it for at least one-third of that time already.

In the meantime, industry will need a lot more service out of that old workhorse — waterflooding.

Good-bye, Henry

Churly McCaustic **May 9, 1983**

I haven't appeared in these pages since Mar. 25, 1968, so I must reintroduce myself. I'm Churly McCaustic.

I've been absent since Henry D. Ralph retired as chief editorial writer of Oil & Gas Journal. I'm returning to say good-bye to Henry, who died Apr. 30 in Frankfurt, Germany.

Henry invented me, just as he invented this column in 1948. He used me — as what he called his "insulting economist" — to convey with humor his often pointed perceptions of the oil industry. In doing so, he set a tone and direction for this column that will endure long after his death. With different approaches and in varying ways, he did the same for the Journal's weekly editorial and news section.

But it's probably for me and the Journally Speaking column that Henry will be remembered.

He used to portray me flailing the air with a ruler, my feet propped on my desk, scolding oil companies for short-sighted refinery management, failure to devise an industry-wide gas price system, or poor public relations. For that, I earned more than my share of scalding letters, including one from "a former reader" who called me something very unflattering. Henry kept that missive on his wall until he retired.

To be honest, I came along fairly late in Henry's career, appearing first on Feb. 10, 1964. He had done quite well without me for more than 15 years, writing nearly every Journally Speaking column and editorial during that period and earning the respect of Journal readers worldwide.

Journal Editor Gene T. Kinney remembers Henry this way: "Henry always did his homework. Although he gained fame as a columnist and a thought leader, his strength lay in the research he did before sitting down to write his columns and editorials."

Henry, Kinney says, "never stopped being a reporter." And he never looked back. "He always knew that the good old days were still to come." He devoted his last column not to reminiscing but to describing the challenges that he felt lay ahead for oilmen.

In the column before that, Henry bade me farewell and through me summarized the journalistic approach that served him so well for more than 30 years. "I like a simple answer to a simple question," he had me say. "I try to analyze a problem and get down to its fundamentals."

Then I said, "The petroleum industry could make use of a philosophical approach more often than it does. Philosophy, you know, means love of wisdom, and at times the industry hasn't displayed an abundance of wisdom . . .

"Everybody is so preoccupied with his own immediate business problems that he doesn't take time to look into underlying causes and effects in the industry as a whole.

"So what the industry needs is a few philosophers, if you want to call them that. At least that's my philosophy."

The petroleum industry has lost a philosopher. Good-bye, Henry. And good-bye, Journal readers.

Energy and the media

Bob Tippee **July 11, 1983**

There's one main weakness in the Media Institute's new book, "Energy Coverage — Media Panic."

Like most press criticism, it appears too late.

In the popular press, energy isn't big news anymore. So "Energy Coverage" may do more to sharpen the defenses of those who consider themselves potential victims of media panic than to prevent recurrences of the journalistic lapses that characterized coverage of the last two energy crises.

The book is a collection of papers and discussion transcripts from a conference in England of energy communicators, oil firm executives, and others involved in oil and nuclear power issues. Thus, it provides a smattering of insights into international media treatment of the oil price jumps of 1973-74 and 1979-80. And it devoted much attention to how the popular press has handled the U.S. oil and gas industry.

An important theme of the book is the media's susceptibility to the "big conspiracy theory" — that major oil companies conspired to control oil supplies and drive up prices.

That susceptibility sprang in part from the investigative journalism craze lingering from the Watergate scandal. It also resulted from the lack of energy expertise among newspeople, which steered coverage away from the technological and economic aspects of the story in favor of the political and personal.

Resulting distortions tailored energy policy. Says former television newsman Robert S. Goralski, now with Gulf Oil Corp., "Great was the day when we discovered a family of six living in a Victorian house, uninsulated, and they couldn't afford to turn on the lights or the television, and they were sitting there in the dark. This was a particularly good story. Everybody loved it. High impact.

"And you see this kind of thing in the programs every night — big, complex stories are reduced to the plight of the family. It's bound to have an effect on somebody in Washington who has to legislate some way to get that television set back on for that family."

Apparently so. Herbert Schmertz, Mobil Oil Corp. vice-president of public affairs, tells of visiting nearly every U.S. senator while the "windfall profits" tax on oil was being debated. "The vast majority" of them, he says, told him, "Yes, I understand why this is bad legislation, but there's absolutely no way, considering how it is appearing in the press, that I can go back to my constituents without doing some damage to the oil companies. When you can get public opinion back on your side, come back and see me."

It's too bad law and news must be made this way. In its watchdog role, the press should help keep bad laws off the books. But in the case of U.S. oil and gas law, it seems to have worked the reverse.

There are signs of improvement. Natural gas price decontrol has received support from some surprising quarters recently.

Perhaps the petroleum market shifts since 1973 have convinced some reporters and editors that the law of supply and demand is something more than capitalistic rhetoric.

Or maybe there simply are more interesting conspiracies in other fields these days.

The road to recovery

Jim West **October 31, 1983**

Anyone who doesn't know that U.S. independent operators are taking it on the chin during the industry's recession hasn't been keeping up with the news — the bad news, that is.

Independents' profits generally are down much more sharply than those of large, integrated companies, many independents are operating under protection of Chapter 11 of the bankruptcy code, and most who ventured into contract drilling are in serious financial trouble because of a plunge in demand for rigs that knocked the props out from under footage and day rates.

Granted, much of the news is bad. But not all of it fits that category. In fact, some of the most recent news shows that the worst of the recession is over, and the industry is on the road to recovery.

That's the view of Jon Rex Jones, Houston. Featured on our cover, Jones is the incoming president of the Independent Petroleum Association of America. He'll take office this week when IPAA members gather in New Orleans for their annual meeting.

"We are on a gradual but steady course to recovery," Jones declares.

As evidence he cites a rising count of active rotary rigs after the tally bottomed out last April, a reversal of the decline in active seismograph crews, and stabilization of crude oil prices.

Among the good news for independent producers is an improvement in the efficiency of drilling crews, estimated by Jones at 20% — even 50% in some cases.

"We can pick and choose from the most efficient and newest rigs, and we are paying bargain prices for some oil field supplies," Jones says.

That acknowledges the troubles that linger for the service/supply segment of the industry.

An example: Drilling contractors, like producers, can reduce staff and cut other expenses during industry slumps. But the basic tool of contractors, the drilling rig, can't be laid off. Jones says, "It just hangs around, whether or not of any value, waiting to be put to work." He doesn't expect "a tremendous increase" in contractors' rates until the U.S. active rig count climbs to 3,000 or more.

Producers have persistent troubles, too. One of the biggest is federal control of natural gas wellhead prices.

Jones says, "For 30 years natural gas prices have been under government regulation. Now, when the political environment could be favorable to eliminate those controls, our industry is caught in the throes of its own political battle.

"Natural gas producers are polarized by 'old' gas vs. 'new' gas debates. We have some producers with 'good' contracts they fear will be abrogated and others with 'bad' contracts they are trying to abrogate."

In his new post, Jones will try to help map a trail through all sorts of minefields surrounding issues such as natural gas decontrol and the possibility of additional taxes on oil and gas.

No doubt about it, the next couple of years will be challenging ones for Jon Rex Jones. We wish him luck.

A time for optimism

Rick Hagar **January 2, 1984**

I heard Ken Lay, president of Transco Energy Co. tell an interesting story the other day.

It seems there was a commercial airline pilot who, over cocktails, confided to a friend that he had a serious vision problem.

"I've never told this to anyone before," the pilot said, "but I have no depth perception."

"Goodness," his friend replied. "How do you manage to land your jet?"

The pilot claimed he had no problem finding the runway. The problem came when he was about to touch down. He couldn't tell when his landing gear would touch the concrete.

"Isn't that dangerous?" his friend asked.

"No, not really. I just ease her down slowly, you see. Very slowly. And when I hear my copilot scream, 'WHAT'RE YOU DOING?!', I pull the nose up and land."

Some of the oilmen in the crowd laughed. Others, particularly those with companies still trying to make that safe landing, were lost in thought.

It's been a rough ride for most in the oil and gas business, particularly during the year just passed.

Did we reach bottom last year? Or do we still need to be listening for that copilot's scream?

I'm optimistic about 1984.

More than 2,800 rigs were at work as 1983 came to a close. Although that's nowhere near the 4,500 turning to the right at the end of 1981, it's 1,000 more than the number drilling at one point last spring.

Maybe oil prices haven't climbed to $60/bbl, and deep gas isn't bringing $10/Mcf anymore. But can't a person make a good living at $29/bbl oil and $2.80/Mcf gas with today's drilling economics?

Pipelines won't take your gas, you say? Well, I keep hearing reports of producers boosting takes by finding alternate markets — if they're willing to roll back the wellhead price.

And isn't there a chill in the air? Maybe someone will turn up his thermostat a notch and use a bit more gas.

It was big news last month around OGJ's Houston office — particularly among the advertising salesmen — when a service company announced it is coming out of bankruptcy with a plan to repay all its creditors.

Can some of the bankrupt independents be far behind?

Drilling may fall off during this year's first quarter, but that's only normal for this time of year. The big question is: How strong will it rebound in the spring?

I'm impressed by the attitude of Chevron's L. C. Soileau, senior vice-president for exploration, land, and production, who says, "For companies with the money to do it, now is the time to drill."

Bananaol?

G. Alan Petzet **February 20, 1984**

Oil & Gas Journal's popular refining, pipeline, construction, and other surveys are the province of Aileen Cantrell.

Cantrell, OGJ's director of editorial surveys, has been handling these reports for 12 years.

To her credit, she is a stickler for accuracy. That's saying a lot with the volume of information she has to compile.

She sends questionnaires to be completed and returned and some come back with cover letters from the respondents.

Oil is a particularly proud industry, and a company that shuts down a refinery or postpones a big capital project is often reluctant to see that fact in print.

That pushes Cantrell into a lot of interesting followup action to convince the respondents, many of whom by now are good "telephone friends," to release information.

A lot of people, as we might say in the U.S., would have gone bananas after 12 years of that. Not Cantrell, but you might not know it to see some of the return correspondence.

For example, one company returned a blank survey form with the following cover letter:

"Our company wishes not to be included in the 1984 survey or in any future surveys.

"We consider all information concerning our company to be confidential, and we therefore require that no mention of this company be included in your surveys without written consent.

"If you decide to ignore this request, please notify me so that I might take the proper action to influence your decision."

That one was signed by the vice-president of refining, who forwarded a carbon copy to legal counsel.

Soothing some of the frayed nerves resulting from letters like that are replies like the following one from Hawaii:

"The annual U.S. refining survey has been completed and is attached. We would appreciate a reprint of the survey when available.

"There was one area of production we were unable to include. We produce 1,000 lb/year of bananas. We would be interested in how our efforts in this category stack up relative to other respondents."

That letter helped conjure visions of unleaded gasoline with bananol, entitlements to help refiners with access only to higher cost new bananas compete with those who control lower cost old bananas and, of course, bananol tankwagon prices.

It was too late to survey other possible banana-producing refineries this year, Cantrell says.

Nevertheless, OGJ's refining survey next month (Mar. 26) should be top banana.

The lesson of Mukluk

Bob Williams **June 25, 1984**

North American Arctic oil and gas operators can learn much from the lesson of mukluk.

The Eskimos' caribou skin boot, that is. Mukluk with a capital M spells history's costliest dry hole and perhaps most publicized wildcat of the modern petroleum era. More about that later.

For centuries, native peoples of the North American Arctic have relied on caribou to supply much of their daily subsistence: meat, tallow, skin, bone, antlers, hooves, and the like yielded food, clothing, shelter, tools, weapons, and ornaments.

In the barren Arctic, natives since prehistory have had to wring every possible use from this stately animal to survive. Not out of choice. The bounty was there, but first had to come the wisdom to see it. And then the care and efficiency to use the resource wisely to preserve it for the times when the caribou might be scarce.

That spirit persists among oil and gas operators today in the North American Arctic, especially in Alaska. Alaska Airlines is flying energy writers from around the U.S. to the Frontier State in an effort to update them on the next big oil and gas story there. It might also serve to offset the mass media delusion that the future of Alaskan oil after Mukluk is a writeoff.

Prudhoemania persists, the $2 billion Mukluk failure notwithstanding. Industry still will probe for supergiants in the U.S. and Canadian sectors of the Beaufort Sea. The big exploration story currently, however, is the first Bering Sea campaign.

But the big Arctic petroleum story carries a low profile, typified by the quiet, methodical approach ARCO, Conoco, et al., will use to tap a Greater Prudhoe area resource base pegged at more than 60 billion bbl of tough-to-produce oil.

Industry will need an Eskimo's kind of wisdom, care, and efficiency to exploit and preserve this third generation of North Slope reserves for the next supply short times. And that story highlights this year's North American Arctic Report.

That holds another lesson for whoever will be formulating Canada's new energy policy and for the Alaskan government officials facing a possible huge revenue shortfall with declines at Prudhoe Bay and Kuparuk fields next decade: Oil and gas firms can be "caribou," but not if they become "golden gooses" killed by onerous taxes and regulations.

Sometimes another Prudhoe can be right under our noses if we follow the lesson of the Eskimo's caribou skin boot.

Dividends from the past

Roger Vielvoye **September 24, 1984**

Linking North Africa to southern Europe with a deepwater gas pipeline is, as our report on the Mediterranean region in this issue shows, a tremendous technological achievement that will have far-reaching political and gas supply implications.

Laying the first section of the 1,551 mile line through the deserts of Algeria and Tunisia and then across the Mediterranean seabed at depths of more than 1,900 ft required technical skill and political persistence.

But when the line came ashore on the Italian mainland after a landfall in Sicily, SNAM SpA, the ENI company masterminding the project, ran into a new set of problems.

From the toe of Italy to Minerbio in the north, where the final phase of construction is nearing completion, the route ran through 20,000 properties, including farms, vineyards, and sites of archeological interest.

Detailed route surveys by Saipem, the ENI company responsible for pipelaying, identified the areas of archeological interest. Those were avoided.

Inevitably, the trenching work turned up ancient sites that were previously unknown, particularly in the Latium area near Rome.

Whenever pipeliners hit one of these new sites, the local Inspectorate of Archeology was immediately summoned, and SNAM provided men and resources to excavate the site.

In the Latium area alone, SNAM discovered burial sites and a Roman villa near Pontecorvo, southeast of Frosinone, and another villa and Roman statue near San Polo Dei Cavalieri. At Palombara Sabina digging uncovered a pre-Roman burial chamber containing the remains of ancient Sabine warriors with iron swords and shields by their sides.

The Sabines eventually were conquered by Rome in the third century B.C. and are now more popularly remembered for the rape of their women by the Romans.

The site also yielded something far older: the fossilized remains of a Neanderthal man found among fragments of mammoth tusks and deer bones. After detailed work to date the find, the entire block of rock containing the remains was cut out, crated, and sent to a laboratory for further work.

Not all the sites discovered were in Latium.

Near Pergine Valdarno in Tuscany, excavations were carried out by hand by a specialist firm to unearth ancient walls. And not everything came from the Roman or pre-Roman eras. At Montelucci, also in Tuscany, a small hilltop castle from the dark ages of the 10th century created considerable interest among local archeologists.

Insuring that history was not lost and that the impact of the pipeline project on communities along the route was kept to a minimum was costly and time consuming. But it is the kind of investment that will pay dividends for ENI in the future.

Robert J. Enright

Bob Tippee **November 5, 1984**

Living up to the strict standards, rich traditions, and sharply focused editorial mission of Oil & Gas Journal is a challenge. Making a difference is something special.

The Journal and its readers lost one of those special few individuals who made a big difference when Robert J. Enright, executive editor, died suddenly Oct. 24 at his home in Tulsa. He was 59.

In his 34 years as a Journal editor, Bob covered the petroleum story from the plains of Texas, to the boardrooms of New York, to the deserts of the Middle East. He spent 11 years in the master juggler's position of news editor. From 1977 until his death, he guided editorial policy, choreographed special reports, planned cover photographs, and fulfilled an immeasurable array of senior staff responsibilities as executive editor.

All along the way, he made a difference.

He made a difference in the lives of his coworkers with his casual, Kansas style. A tornado of missed deadlines and missing color separations couldn't quiet the happy but unidentifiable tune he always whistled or suppress his characteristic greeting: "Hello, ol' pro."

He made a difference in this magazine. How the Journal reports the news, how it looks, how it views the industry it covers — all reflect Bob's eye for detail and boundless knowledge.

And he made a difference in how the readers he served look at their industry.

Perhaps it was his clear, Kansas eye that enabled him to get to the heart of things so well.

Who can forget how Bob straightened the record when oil and gas discoveries in Mexico triggered reports that the country had 200 billion bbl of oil reserves and thus might become another Saudi Arabia?

In this column on Jan. 1, 1979, he pointed out that the reported "reserves" figure was Mexico's estimate for potential resources of crude oil, natural gas, and natural gas liquids. He cited the Petroleos Mexicanos oil, gas, and gas liquids reserves estimate of 20 billion bbl of oil equivalent and repeated the Journal oil reserves estimate, about 16 billion bbl.

Then he noted that Saudi Arabia had increased its reserves by about that amount the previous year and asked, "Does this mean Saudi Arabia has added another Mexico?"

It will be quite some time before the Journal staff begins to comprehend the extent of its loss.

And it's with some relief that we have only to write this and not to say it out loud: So long, ol' pro.

Correctiing misteaks

Jim West **December 10, 1984**

It is an enduring source of grief to editors that all writing stands a good chance of having at least two things in common.

Lurking somewhere in news stories, technical articles, quarterly earnings reports, annual reports — any written material — there likely is a typographical error. And there is an equally good chance that no matter how many proofreaders check the material, no one will catch all "typos" before publication.

We've certainly made our share of typographical and other mistakes — things like reporting storage capacity in barrels a day instead of plain old barrels, listing the Beaufort Sea open water drilling season as beginning in November instead of ending in November, and placing Colorado's Rangely field in Piceance County.

A reader caught that last error with a letter pointing out that Rangely field is in Garfield County, Colo. "There is no Piceance County anywhere," he wrote. Quickly following the first letter was his second one: "Correction, please. Rangely field is in Rio Blanco County, not Garfield County. My face is slightly red." All writers know the feeling.

We play turnabout in many cases and catch errors in press releases.

A Houston firm once reported its area of operations includes the U.S. Overburst Belt. We changed that to Overthrust. A Calgary drilling contractor announced formation of a joint venture under which it planned to sire and train northern Canada natives. Hire, not sire, was the intended word, of course.

We also have been notified that sulfur produced at a Rocky Mountain gas processing plant will be transported via tank cars in molten liquid form. Molten liquid? As opposed to solid liquid?

Another press release carried the news that the first conference on a certain energy subject was to be held again in Boulder, Colo. First? Again? Something's wrong somewhere.

We received a geological report from an Appalachian basin state saying,

"Sinkholes were being found in this valley long before man set foot there." How can anyone know what was being found anywhere before man set foot there?

Early last October we received this report from the U.S. Department of Energy: "A major thrust of DOE is to investigate how to use the nation's abundant coal reserves to alleviate the shortage of transportation fuels." Lots of refiners would be surprised to hear of such a sharp reversal in products supply/demand balance.

Some writing — and speaking — isn't erroneous. It's just overdone.

For example, a spokesman for a Canadian oil company once told an agricultural group, "Farmers can look forward to lots of supply and therefore lots of availability of petroleum based products for the next 2-3 years. There will be limited demand and therefore soft markets for fuels." No one can argue with that line of reasoning.

Although everyone tries, it's very hard to catch every error in everything that's published. To prove that point, try proofreading: "Thiss this sentence contains four erorrs." If you found only three errors and corrected them, you just published the fourth error.

Grace vs. overspending

Bob Tippee **December 24, 1984**

There are two ways to cover a speech, neither of which works when J. Peter Grace talks about government spending.

One way is to take notes during the speech. But the mouth is quicker than the hand under normal conditions. When the chairman of W. R. Grace & Co. discusses findings of the President's Private Sector Survey on Cost Control, information flies by so rapidly the hand quickly drops out of the race.

The other way is to record the speech on tape and transcribe it. That might work with Grace. Even then, however, it would take a team of statisticians to transcribe his dizzying bursts of data on government overspending, inefficiency, and unaccountability.

Grace discussed the cost control study, of which he was chairman, at Rocky Mountain Oil and Gas Association's annual meeting this fall. He talked for 45 min and seemed to be just getting warmed up when he quit, thanked the audience, and departed — off, no doubt, for another of the 100 or so such speeches he delivered this year.

One observer at the Rmoga meeting said something like, "If I were his public affairs adviser, I'd tell him to slow down and use fewer numbers. A listener can't absorb all that."

But isn't that the point? The way government spends money is mind-boggling. Two or three or a dozen figures don't tell the story. A torrent of figures may obscure individual bits of information, but it does better at communicating the proportions of the problem.

Grace's study group, which has come to be called the Grace Commission, found 2,478 ways to reduce federal waste, inefficiency, and overspending. It said its proposals could save $424.4 billion during 3 years.

Grace apparently can recite the amount for each.

In fact, his command of his subject forms a credible base for some refreshingly bold statements about the federal government, which Grace says spends $1.25 "for every dollar we take in."

If Americans keep voting for essentially the same 535 congressmen, he declares, "they'll keep doing this to us."

There's $850 billion owed the U.S. government, $38 billion of it in default, none of it "aged," and all of it kept track of by 332 incompatible accounting systems.

"It's your money, and they don't care," Grace declares. "Everything is wasted, and it's because no efficient business principles apply. When our children wake up and see what we did to them, will they ever forgive us?" The solution? "Scare any big spender."

Those are fighting words in parts of Washington, D.C. And the fight will heat up when Citizens Against Waste, an effort by Grace and columnist Jack Anderson, presents a petition against wasteful spending to lawmakers next month.

Grace will be called all kinds of things.

Here's hoping they don't leave out the title due citizens who devote time, energy, and money to identifying important national problems and promoting reasonable solutions: statesman.

Offshore preparedness

Rick Hagar **January 21, 1985**

The view from the ballast control room's porthole told the story.

As the big semisubmersible drilling rig pitched to starboard in the storm, only ocean could be seen. But the horizon dropped to the midway point of the porthole as the rig rolled to port.

The wind was 70 knots. Seas were 50 ft. And the Victory class semisubmersible, owned by Ocean Drilling & Exploration Co., was rolling 2 degrees to port, 2.5 degrees to starboard.

Odeco's Robert E. Elmwood began switching levers on the control console before him.

"The goal is to adjust the ballast to keep the rig perfectly stable," he explained. "While drilling, we should never be rolling this much."

Soft hisses signaled the opening of a series of valves he controlled.

But before the operation was complete, an alarm warned that the port bow tank in the semisubmersible's center hull was flooding. A work boat apparently had slammed against the rig, gouging a hole in the tank.

Elmwood quickly sprang into action, activating another series of valves and pumps to counteract the rig's tendency to list to port.

As the port list increased to 5 degrees, Elmwood considered using the emergency deballast system. However, that option wasn't needed as pumps took hold and the list slowly became less pronounced.

Elmwood then stood back, threw open curtains, and walked out of the ballast control room onto the ground floor of Odeco's building on Canal Street in New Orleans.

Elmwood, Odeco's senior training instructor, had just taken his guest through one of the dozen scenarios his company's unique ballast control room simulator can duplicate.

Although the room overlooks Interstate 10, its design makes it easy to imagine 500 ft of water beneath the floor, miles from the nearest coastline.

The simulator was added to Odeco's training curriculum following the 1982 loss with all hands of the company's Ocean Ranger semisubmersible off Canada.

The U.S. Coast Guard said the sinking stemmed from a chain of events that began after a broken porthole in the ballast control room allowed seawater inside.

The sinking claimed 84 lives.

Odeco, with the simulator and a $6 million/year safety program, wants to be sure a similar disaster never happens again.

The shellfish conspiracy

Bob Williams **January 14, 1985**

The most advanced form of communication in California is the bumper sticker.

That figures in a state (of mind?) where announcing that one is autoless can elicit sagged jaws and sometimes the sign of the cross.

You can see any message on chrome in Los Angeles from the trite — Welcome to California/Now go home — to the seriously weird — Nuke gay whales. A tailgate aphorism must be truly different to be noticed in Lotusland.

So it took a keen reporter's eye to pick up this one on the back of a sleek, unpronounceably Italian, 6 mpg vehicle: Stop Big Oil exploitation in the

Santa Barbara Channel — Kick Oil Out Constituency (KOOC).

How could an OGJ West Coast editor resist pulling alongside for a rush hour dialogue with the driver?

OGJ: Thought I'd heard of all the environmental groups. What's your beef against drilling for oil in the channel?

KOOC: Oil? Who cares about oil drilling? We're opposed to Big Oil resource exploitation that's much more insidious — part of a vast cosmic conspiracy that started in the Gulf of Mexico. Shellfish.

OGJ: Say what?

KOOC: Believe it. Phillips Petroleum Co. is the ringleader here. Started when Phillips had trouble with mussel buildup on the legs of its Platforms Hogan and Houchin in Carpinteria field. Mussels got so thick, it worried Phillips that the critters would play hob with the platform's resistance to ocean currents. They were cleaning the mussels off routinely when some wiseguy figures there might be money in it. So they contracted out the mussel harvesting to a Santa Barbara firm. That way, clearing the platform legs pays for itself, the local restaurants have a temporary deliverability surplus of mussels, and an entrepeneur is cleaning up.

OGJ: So what's the problem?

KOOC: Can't you see? Sure, it's starting small. That's how it began in the Gulf of Mexico, and now look — record catches of all kinds of commercial fish because of those platforms. And then what? Throwing open all U.S. waters to rapacious fish harvesting from platforms? All that marine life around platforms will attract predators — and no provisions for a Shark Impact Report for surfers! We don't need the extra fish. It would be better to conserve the seafood at hand. At best, one platform could supply only 1 week's mussel demand at, say, the Hungry Tiger restaurant chain. Better to subsidize alternate foods, like imported truffles, or homegrown synthetic food, like tofu.

OGJ: Ugh. Who's going to pay for that? Californians, who consume more seafood than anyone else?

KOOC: Don't tell me about seafood. I own a commercial fishing fleet. (Gunning the engine) I figured you might be one of those fools who don't see the truth — that the oil industry's sole motivation is to boost already obscene windfall seafood profits!

Completion problems

Jim Leonard **April 1, 1985**

A technical editor's job is quite different from, say, that of an oil company production engineer.

I did the latter for many years before becoming OGJ production editor.

And yet, putting together the annual production report in this issue reminded me that the two professions have a common thread.

Whether completing a well or completing a special report, almost anything can happen.

The problems we'd like to do without in putting together an annual Production Report are fresh in my mind. But some of the mildly frustrating ones we experienced as a production engineer are almost as fresh, even after all these years.

For instance:

1. The service company called me at home and said its crew was rigged up and ready to perforate. (I overslept and the well was 65 caliche-road miles away.)

2. Hadn't slept for 30 hr. Tools were at 12,000 ft and the collar locator on the correlation log kept shorting out. (Strange, it worked like a jewel on the previous 84 runs over in Ector County).

3. The Amerada bomb was run to total depth (9,812 ft) for a bottomhole pressure. Weevil closed the master valve on the wire line for a shutin pressure. (You closed whaaaaaat?)

4. While swabbing salt water from a dead (?) gas well, it kicked off with a wild blow, the lubricator broke off at the hammer union, and the swab blew out of the hole from 3,000 ft. (Whose turn is it to close the master valve? We are taking turns, aren't we?)

5. In any large field development, it is always necessary to go through seven gates and cattle guards to test the least-productive well. (And seven coming back, unless you find a shortcut.)

But bad days aside, most wells were eventually completed. And after the publishing equivalent of some of the oil field problems sampled above, the Journal's annual Production Report has been completed.

We think the results justified the effort. We hope you agree.

Thanks, Daimler and Benz

John L. Kennedy **April 29, 1985**

Sure, demand for oil products is way down from, say, 1979. And gasoline demand in the U.S., for instance, is about where it was 10 years ago.

Gasoline is more than plentiful, and refiners hope that before too long, margins will increase to zero.

It's tough. But it could have been worse were it not for an event that took place 100 years ago.

In Germany in 1885, Gottlieb Daimler and Karl Benz, experimenting separately, each developed a gasoline-powered internal combustion engine to power a vehicle to carry passengers.

It has been a durable concept.

And it deserves at least a moment of recognition during this 100th anniversary year by those of us whose livelihood is affected in part by the consumption of gasoline.

Forget the social changes the automobile brought. Forget the millions of jobs the industry has provided around the world. Forget its contributions to economies during the past century.

And forget for a moment the cost of making unleaded gasoline, meeting refinery effluent regs, today's surplus, and other problems.

Instead, think about what gasoline demand would be today if it weren't for the gasoline-powered, internal-combustion, passenger carrying vehicle.

It wouldn't be 6.7 million b/d in the U.S. alone.

And if the electric automobile, developed only a few years later, had prevailed, the Age of Illumination would have yielded to the Fuel Oil Age. The Gasoline Age would never have happened.

Other developments have come along which have lightened the hearts of refiners. Assembly-line production and increased standards of living meant tens of millions of people could buy cars during the last 100 years. Big engines, air conditioning, power assists, automatic transmissions, and other goodies came along and sucked up gasoline.

But, of course, it's the concept that started it all. And 100 years later, there is no evidence that a significant portion of those billions of cubic inches of displacement around the world that must be filled with gasoline every day will be gone soon.

Certainly, those making and selling gasoline would have made a living making and selling something else were it not for the automobile.

And if Daimler and Benz had not developed the basic engine configuration that is still used in most automobiles, someone else might well have.

But they did, and someone else didn't.

So thanks, Daimler and Benz.

Tunnels to Troll?

Roger Vielvoye **June 10, 1985**

The scene is a platform in Phillips Petroleum Norway's main Ekofisk complex in the Norwegian North Sea.

A hazy sun is setting over a couple of veteran Phillips hands making a return visit to the Ekofisk cluster of fields after many years' absence.

They lean on a guard rail and gaze at the concrete breakwater that surrounds Ekofisk Center. One of the hands subconsciously starts to count the rows of perforations above the level of the waves.

Suddenly he looks puzzled. Is his memory playing tricks? There are only four lines of "port holes" in the concrete instead of the five he remembered. Nothing is wrong with your memory, confirms his colleague. There definitely is a row missing.

Together they consider possible causes. Subsidence is unlikely because that usually occurs only where the reservoir is less than 6,500 ft below the seabed. Ekofisk reservoir is below 10,000 ft.

Other reasons are reviewed and ruled out. Eventually they come back to subsidence as the only possible explanation. Management is alerted. Checks are carried out. It is discovered that the seabed around seven of the platforms is slowly sinking.

A nice story but unfortunately some way from the truth — just one of the myths that are created almost every day by crews operating in the North Sea.

As the Northwest Europe report shows, Ekofisk is suffering from severe subsidence. However, the problem was not discovered while leaning over a rail contemplating the sunset but through painstaking collation of many observations, mainly of wave action on the lower decks of steel platforms.

Phillips is assessing several ways to protect the platforms, including one with a science fiction touch.

Norwegian scientists are trying to perfect a series of floating structures that would concentrate wave energy to allow more efficient harvesting of power from the sea. Reversal of the system so wave power is dissipated and not enhanced might result in an ideal floating barrier for vulnerable Ekofisk platforms.

There are other stories that sound as though they should be a myth but have a basis in fact.

Europe's biggest offshore gas field, Troll, has twin problems: deep water and a very shallow reservoir that will be impossible to tap with directional holes from a single giant platform.

One proposed solution: Drive three tunnels 50 miles from the Norwegian mainland under the seabed to Troll, then blast a series of caverns for drilling/production facilities.

Tunneling to Troll may sound fictional — and more than a little dangerous — but it was put forward as a serious proposition by a group of Norwegian tunnel constructors. Statoil has even run a feasibility study. Technically, the idea is sound. With modern technology the tunnels and caverns could be constructed.

But for all those concerned about becoming troglodytes rather than platform dwellers there is good news. Even taking a generous view of future European gas prices, tunnels and caverns are too expensive.

Digging up the turf

Rick Hagar **July 1, 1985**

Oil & Gas Journal takes a look in this issue at changes that are gripping the U.S. natural gas industry.

The biggest change, of course, is the new emphasis on marketing.

I have only to step out my front door to see evidence of that. The gas distribution company in Southwest Houston has dug up portions of my prized St. Augustine grass lawn.

The distributor is expanding sales with a campaign that has neighbors scrambling to convert electric space furnaces and water heaters.

The distributor's offer is to furnish and install a high efficiency, 40 gal gas water heater, a high efficiency gas furnace, 100 ft of line, and the necessary permits and inspections — all for the cost of the furnace.

But that's not all. The distributor also will finance the package.

The offer is so attractive that some neighbors were incredulous.

They compared 1 year of all-electric service with 1 year of combination gas and electric service and found that gas — even with conversion and finance costs figured in — could save them as much as $20/month under current rate structures. Wait, warned a neighbor who happens to be a lawyer for the local electric utility. Electric heat pumps are the way to go, he said.

The gas distributor, however, had done its marketing job well. The entire block of neighbors, with the exception of the lawyer and a Gulf Oil Corp. manager transferred to San Francisco, chose to convert to gas.

This conversion hurt the prized St. Augustine grass because the nearest gas line is under my yard.

The experience brought home a reality of the new gas market: Industry is learning how to compete and is trying to build market share with innovations throughout the U.S.

The way of doing business is undergoing long lasting changes. Producers, transporters, and distributors who cling to old ways will find themselves losing market share.

It's best to pay attention to what's happening. It might be your turf that's dug up next time.

Back to work with brine

Michael Obel **August 12, 1985**

It's no secret that the number of petroleum geologists and other personnel employed by oil and gas companies is declining.

Large mergers and a general contraction of the industry are responsible.

The American Association of Petroleum Geologists has been tracking the effect on its members from three of the largest mergers: Mobil Corp.-Superior Oil Co., Chevron Corp.-Gulf Corp., and Texaco Inc.-Getty Oil Co.

Before the three mergers, AAPG had 3,343 members employed by those six firms. Today the association's records show that 521, or about 16%, have left for a variety of reasons.

"Our data indicate that most have managed to find employment," says Fred Dix, AAPG executive director. "But I don't think this is over. It hasn't completely shaken out."

Early retirement is one of the reasons there are fewer geologists employed by companies today than in the early 1980s. It's a widely used approach to pare personnel costs.

The Conference Board, New York, reports that of 363 companies it surveyed recently, more than 60% had early retirement inducements in their pension plans.

Phillips Petroleum Co. may be doing as much as any major U.S. oil and gas company to encourage early retirement.

Earlier this year Phillips offered early retirement incentives to about 2,400 employees due to reach 58 years of age in 1985. In the past 3 months, 1,930 employees accepted the offer.

Now Phillips has issued a second offer, this time to employees reaching the ages of 55-57 in 1985.

For one reason or another, the ranks of employees in the U.S. petroleum industry continue to dwindle. There must be some way to put all those people back to work.

There is, says Don Vernau, president of the Ohio Oil & Gas Association. Simply start a government campaign similar to the Agriculture Department's program in which farmers are paid for not planting crops.

Vernau jokingly suggests that the Environmental Protection Agency pay oilmen for not producing brine in excess of 50,000 b/d or so.

Think of the demand for personnel that could create.

Geophysicists and petroleum geologists could help producers find brine reserves to not produce. Reservoir engineers would be required to determine the maximum efficient nonrate of production to assure that brine reservoir nondepletion is conducted consistent with accepted engineering practice.

The ripple effect would put untold numbers of administrative personnel back to work.

And if EPA paid its subsidy on a sliding scale, depending on lead or mercury content, investors would insist on brine assays before making a decision. That could spark demand for chemists.

But perhaps there is a better idea. What about dropping threats to increase industry taxes, ending the moratoriums on offshore leasing, and junking controls on natural gas prices and markets?

A firm company, maybe?

Bob Tippee **September 9, 1985**

The trouble with new staffers is there's always one who knows something you don't.

Oil & Gas Journal has one like that. He's Staff Writer Mike Obel, not quite a 1 year veteran with OGJ.

Not long ago he asked, ever so innocently, whether there wasn't a difference between the meanings of "company" and "firm," which he noticed often were used synonymously in Journal stories.

He remembered some professor somewhere who insisted the terms weren't interchangeable.

Well, you know how professors are, Mike was told. What do they know?

Come to find out — in Webster's Third New International Dictionary when Mike wasn't looking — "company" and "firm" don't mean the same thing.

Technicalities aside, it boils down to Exxon is a company but not a firm, which is more like an association of lawyers.

This distinction is trifling unless you have to write the Oil & Gas Journal 400 story. No writer wants to belabor his prose with unnecessarily repeated words, even in a story that's mostly numbers.

So what do you do when you're writing about the 400 biggest, public oil and gas producers in the U.S. and you discover you can't use your former favorite synonym for "company"?

You get out the thesaurus. There, you discover the language is pitiably deficient.

"Association"? Too vague. "Corporation"? A growing number of the — uh — companies on the list are publicly traded partnerships. "Concern"? Weak. "Business"? Maybe. "Outfit"? Colloquial. "Troupe"? "Group"? "Assembly"? No way.

Woe is the writer!

Here's an industry in flux, with companies disappearing left and right and the survivors jumping all over the place in the assets and other rankings, and the poor scribe must proceed without benefit of a reasonable substitute for a term as clumsy as "company."

To be precise, the need for the word — or a synonym — arose 57 times in the writing of the 1985 OGJ 400.

"Firm" never appears, which should make Obel and his confounded professor happy.

The writer opted for "company" or its plural 53 times, indulged "concern" three times, and stooped to "business" once.

You can come up with numbers like that when you have computers, which, with Statistics Editor Glenda Smith at the controls, is how OGJ produces this annual mountain of data.

This is the third time around, and it isn't getting any easier. How could it, with smart-alecky newcomers like Obel stealing all the helpful words?

Next year he'll probably suggest the story contain fewer numbers.

California priorities

Bob Williams **September 16, 1985**

Covering the issue of how much drilling will be allowed off California has been a fascinating, complex job since OCS Sale 53 was conducted May 28, 1981.

Industry has performed superbly in allaying environmental concerns and accenting the benefits of reducing U.S. oil imports and trade deficits, creating jobs, and adding revenues to state and federal coffers. A perfect record has dimmed the specter of the 1969 Santa Barbara Channel oil spill. A logical person would have trouble resisting the weight of pro-drilling arguments.

So Sale 53 set records, spurred one of the hottest wildcat plays ever, and will help double Offshore California's platform population by 2000. There even has been unanimous environmental support for the development plan proposed for the OCS granddaddy, Point Arguello oil field.

In view of that dizzying success, new leasing plans should be a snap, right?

Wait, there's more.

Oil alone accounts for 57% of California's energy demand, compared with an average of 45% for all other states. Total petroleum share of the state's energy mix is 85%. California accounts for 10% of all gasoline use in the U.S. It is the world's third largest gasoline market behind the U.S. and U.S.S.R. California uses about twice as much energy per capita as industrial nations with a comparable standard of living. California must import much of its oil and most of its gas as it siphons more than 3% of world oil demand.

You would think new lease sales would be met with open arms. Would you believe congressional moratoriums? Those bans were replaced with a de facto moratorium in the shape of Interior Sec. Don Hodel's offer of 150 tracts as sole lease prospects off California to 2000. The offer dismissed a Minerals Management Service recommendation and elicited laughter as the mildest response from industry.

At least Californians are happy now, right? N-o-o-o-o.

California news media have been filled with angry protests of residents and opposing resolutions by local governments near the 150 tracts.

One reported squabble hints that the tract selection coincided with the districts of pro-drilling congressmen who weren't consulted in the Hodel deal. Newspaper accounts suggest that the heaviest protests were aimed at leasing off the ultra exclusive Palos Verdes Peninsula and Malibu areas of Los Angeles County, with unconsulted representatives grousing that offshore drilling would mar ocean views enjoyed by residents of expensive homes in those areas.

Now it all makes sense.

The Offshore California leasing controversy doesn't boil down to issues of oil spills, air quality, U.S. energy needs, or gas guzzling. One must remember the supreme law of the land in California: real estate property values.

Send SASE

W. D. Moore III **September 23, 1985**

SASE (meaning self-addressed, stamped envelope) has become a legitimate phrase in certain classified ads.

Lonely hearts trade photos that way, one hears. Small collectors use SASE's to swap lists. Box-top offers abound.

Recently, Oil & Gas Journal has done some SASE business. At this moment, there are three technical packages anyone can have in return for an SASE. All are very specific to a small audience and too cumbersome to publish.

One editor wrote about his results last month. Here are some answers and trivia on SASE mechanics.

Question: Do you make money off these? Why don't you just publish the mail-out?

Answer: We do publish manageable, high-interest addenda. Long ones, or those at the edge of mainstream oil and gas business, we delete to save your reading time.

The Journal spends $200-400/offer (total) if U.S. readers pay postage; $1,500 if not. We gladly pay international postage.

Naturally, we hope that this reader service will be so valuable to a few hundred people that it offsets their subscription fees forever and keeps them loyal fans.

Is metered postage on the SASE okay?

Yes. Half the respondents use meter strips. Of the 30% using stamps, half put on too much postage.

Do we need a cover letter?

Just identify what you want. Many people send a copy of the offer or a stick-on note. Offers are addressed to a title instead of a person for rapid handling.

What happens to those who don't send an SASE?

We usually send the material anyway. But the work discourages future public service offers. Time and money are not easy to volunteer during a recession.

How much of your staff's time does it take?

Staff? That's why we want an SASE.

If we don't have stamps, is it okay to send cash?

So far, your drilling editor has supplemented his salary by 81 cents.

Why do some other offers take 4-6 weeks for delivery?

They work in a large warehouse operated by a large bureaucracy. Our inbound mail gets here 2-5 days after the postmark (4-9 from overseas). Filled requests probably get back in the same time. Our motto: In by 9, out by 9:15.

Who sends for the packages?

All types, mostly engineers. Only 28% of a recent sample are subscribers. Fifty-eight percent read someone else's copy (we think we know whose). Thirteen percent neither subscribe nor have a company sub; no telling how they came by the offer. But all of you are welcome.

Sabine Pass sentinels

Leo R. Aalund **November 11, 1985**

In the green ocean of marshland that makes up much of South Louisiana, any elevation a few feet above sea level is high ground.

You don't need a mountain there to see to the horizon.

A blacktop road, its bed only slightly higher than the marsh, serves nicely as a vantage point for this watery panorama. Generally, beyond this horizon there is more horizon.

There's a stretch like this near Johnson's Bayou, La. Heading west along it, a traveler is puzzled by the first hazy glimpse of huge forms rising from the plain of marsh grass miles away. A mile or 2 down the road the structures slowly emerge as a mighty fleet of offshore rigs, seemingly drilling smack in the middle of the marsh.

Those rigs are anchored for several miles along the waterway that takes the Port Arthur Ship Channel and the Sabine River out into the Gulf of Mexico. Overlooked by the classic, antebellum lighthouse of Sabine Pass, they are stacked, mothballed, and, more simply put, out of work.

Rick Hagar, Oil & Gas Journal Gulf Coast news editor, recently counted 24 there during a helicopter flyby on his way out in the gulf.

Given the fact that the Journal has blanketed the offshore industry with its news and technical coverage for more than a quarter of a century, it's likely most of these rigs were once pictured on our pages or on our cover, perhaps as crisp new units right out of the shipyard or while working at sea, or at least were mentioned in the context of new discoveries or field development, maybe in distant oceans.

And there undoubtedly have been Journal editors on the decks of a few, getting article information or sampling a drink and boiled shrimp with drillers, owners, bankers, and assorted dignitaries during a christening party.

All that could be researched.

But the important question is whether Sabine Pass is something like an elephant graveyard, a place where the big ones have gathered to go to their rest.

With port costs what they are, the rigs won't stay there indefinitely. A few might find work outside drilling, but employment for most will depend on a rise in drilling activity, which in turn depends on improvement in industry economic conditions worldwide.

The rigs appeared almost overnight, a sight to behold, during a string of exploratory successes that generated the best geological prospects in years. So once economic prospects turn around, perhaps the stacked rigs will disappear just as quickly.

How they're restructuring

Gene T. Kinney **November 18, 1985**

Everybody's doing it these days. Restructuring, that is.

But they aren't all doing it the same way.

So it seemed like a good idea to examine several companies with different circumstances and approaches. That's what this special API Report is all about. Top executives tell in interviews how they're reshaping their companies to give stockholders a break today and yet survive and prosper tomorrow.

The subjects run the gamut of industry players:

- The White Knight that "saved" another major just before it was to be taken over by T. Boone Pickens and mortgaged the company to the hilt to do so (Chevron Corp.). Chevron is worrying more about tax reform and the federal budget deficit than about oil and gas prices.
- The major that recapitalized to defeat another Pickens takeover effort, returning $4.5 billion in equity to shareholders and going heavily into debt

to do so (Phillips Petroleum Co.). Phillips says that will mean some lost investment opportunities for the time being.

• The major that's "repositioning," not restructuring, without specific prompting from Boone Pickens, and is pouring money into stock repurchases but is fortunate enough not to have to mortgage its future (Amoco Corp.). Amoco will remain a top explorer and sees better days for refining and marketing.

• The flush major with a North Slope cash machine that restructured ahead of any overt threat, sharply reducing its size, cutting payroll to little more than half of 1981 levels, repurchasing stock, going heavily into debt and reducing exploration, preparing for but not predicting $18 oil (Atlantic Richfield Co.).

ARCO sees itself in a unique position. It has been replacing reserves lately as much through engineering as through exploration. That thrust will continue. ARCO may be cutting exploration, but it's putting more money into hydrocarbon technology than ever before.

• The strong independent that strayed but is returning to its main E&P business and shoring up stock prices with a royalty trust, repurchases, and sale of nonoil assets (Louisiana Land & Exploration Co.).

LL&E sees fewer companies and less money spent on exploration but more opportunity for those who are left. The company that evaluates prospects well and uses funds carefully will make it and make it big.

• The midsized integrated company that is taking some of the usual restructuring steps and forming a master limited partnership while shying away from leveraging up (Diamond Shamrock Corp.).

Diamond Shamrock emphasizes a strong balance sheet to be in position to exploit opportunities. It wonders whether some major competitors haven't taken on too much debt.

Good for whom?

Michael Obel **December 9, 1985**

Lowering Reid vapor pressure of U.S. motor gasoline would be good for America, says General Motors Corp.

Why this unsolicited advice from the big carmaker on how someone else can help? Because it wants to make sure that what is good for America continues to be good for GM.

In the case of possible regulations restricting emissions during fueling, that would mean letting refiners pay the bill for what GM says is good for America.

For the politically war-weary refiner, tired of being the Environmental Protection Agency's preferred instrument for cleaner air, GM's efforts may meet unexpected resistance. The key is the grain farming lobby.

Detroit's observers in Washington tell carmakers that EPA is close to dropping the hammer on hydrocarbon emissions released during fueling. Better refiners spending more to cut the vapor pressure of gasoline to 9 psi — GM reasons — than carmakers spending more to install onboard emission control devices.

To that end GM recently announced a study concluding that making refiners pay to clean up ozone pollution would be "competitive" with making carmakers pay.

"Implementing refueling controls at the service station would take longer, and controlling emissions with onboard equipment would be even slower," says GM. "Although low volatility gasoline might cost about 1% more than present gasoline, it would be comparable with onboard or service station control on a cost effectiveness basis."

If refiners had to counter this alone, they would be facing a daunting task. Oil industry successes on Capitol Hill haven't been as numerous lately as those of U.S. carmakers.

But the auto industry can't hold a political candle to the grain farming lobby, the one that was confident enough to suggest recently that this country needs a strategic ethanol reserve. And therein is the silver lining.

Ethanol in gasoline increases its vapor pressure. So lowering vapor pressure may be bad news for the ethanol interests: farmers, blenders, and jobbers in the farm states. As the awareness of GM's efforts courses through the offices of midwestern and southern congressmen, there is going to be a reaction.

There will be alternate studies. Subcommittees will hold hearings to determine how many more beleaguered grain farmers and their banks will fold if EPA forces lower gasoline vapor pressure.

Why, we may even see a few Hollywood political dilettantes fly into Washington to shed a tear or two on the subject.

The trick for the oil industry, once this gets started, is to stay quiet and shuffle along in the skirts (or overalls) of the grain farmers. Too high a profile by petroleum interests might scare off grain farmers' Hollywood friends.

So refiners must remain invisible. Stifle the urge to laugh out loud, please. And next time one of those auto industry studies on vapor pressure crosses your desk, put a corn farmer (or his congressman) on the route slip.

Wonderland revisited

Elizabeth Drake **December 23, 1985**

Turmoil in the U.S. natural gas industry is the accepted norm.

Would producers and pipelines have difficulty adjusting to a calm industry environment in which changes came less frequently?

In Lewis Carroll's "Alice in Wonderland," the little girl fell into a rabbit hole and landed in a place where she encountered an illogical and backward world. She could not understand how creatures functioned in such an environment. And on her journey home, as she adapted to one change, along came another.

Producers want gas to be treated as a commodity, just like oil.

"In the past, gas was almost wrapped up in some regulatory mysticism that resulted in it being treated more as a service than as a commodity," says Joseph H. Crosby, vice-president of crude oil and natural gas sales, Cities Service Oil & Gas Corp., Tulsa.

Producers say the market will become much more responsive to basic economic considerations if gas is treated under commodity marketing and commodity pricing terms.

The natural gas industry is swamped in regulations. Pounds and pounds of reading material was consumed this year alone by industry executives attempting to decipher the regulations so their companies can play or choose not to play under the rules.

Alice got lost when the directions weren't clear. Pipeline company executives are not taking that chance. Until regulations such as Order 436 by the Federal Energy Regulatory Commission are made more clear, most pipelines are not playing.

Now pipelines have until sometime between Dec. 15 and Mar. 31 to decide whether to participate in the nondiscriminatory transportation program.

"There is no certainty in this business," says Crosby. "None of the players in the gas industry today knows what the rules will be.

"We've been through 2 years of transition. We've been through 2 years of going to seminars about transition in the gas industry, and we're still transitioning and not transcending.

"And every time we think we have reached another plateau where things may settle down there is some radical new concept that we have not dealt with that is going to cause us to look at everything with a whole new set of glasses."

Obviously, different players want different rules. Some want decontrol, others want some regulations. While some see block billing as a bad idea, others would not mind if FERC changes its proposal. Federal regulations that help the industry should be retained and the others should be abolished. The question is which regulations are which.

Carroll wrote, "The Queen had only one way of settling all difficulties, great or small. 'Off with his head,' she said without even looking around."

Producers and pipelines are searching for stability.

William Terpstra, Natural Gas Pipeline Co. of America, says he wants to see a combination of regulation and industry freedom established to permit the industry to function. He's not fussy as to how it's done. He just wants it done.

Even Alice eventually woke up.

Home and heavy oil

Bob Williams **January 6, 1986**

Hermosa Beach, Calif., is a lovely town, befitting its name.

A beach resort town of about 18,000 on Santa Monica Bay, it's slightly yuppified and yet a little scruffy around the edges. But there is a nice balance, so it's not as yuppified as Manhattan Beach to the north or as scruffy as Redondo Beach to the south.

There is a fine view of the Pacific Ocean, Pacific Palisades, and Catalina Island when fog or smog doesn't intervene. Nothing beats a brisk run along the surf to El Segundo, where there always is a tanker in view, supplying the big Chevron Corp. refinery with imported oil.

It's a great place to live, but there are drawbacks.

Printing real estate prices ought to be governed by state obscenity laws. Heavy traffic creates unheard-of parking problems. The city is strapped for revenues and has resorted to a parking policy that would befuddle Kafka. And just let a ticket go unpaid. Penalties accrue at a rate to make a loan shark blanch. Then there is a midnight visit, and there goes your car or probably your firstborn if the car's not handy. And there is a need — but not the money — to hire more police.

One smart move by the city put on last year's ballot a measure to permit drilling on city property to raise revenues. Torrance heavy oil field, a giant that may extend offshore, apparently underlies the city. The measure won.

One dumb move by the city puts on next April's ballot a measure asking voters if they oppose drilling in Santa Monica Bay. Don Hodel must be shaking in his boots. Still, proponents say they want to "get the ball rolling" among other bay cities to block offshore drilling after congressional moratorium efforts collapsed.

That offshore drilling would attempt to tap potential resources pegged very conservatively at several billion barrels of oil, of which three-fourths is expected to be less than 20 degrees gravity. It underscores the significance of this week's roundup of heavy crude and tar sands work.

Some in the oil industry think non-OPEC heavy oil is the only viable alternative to too much reliance on imports and offshore is where to find most of the gunky stuff. They also think governments hold the key to its development, with access, permit, and tax policies.

Meanwhile, Hermosa Beach worries about cop shortages and possible revolts by the citizenry as it tries to lose the title of "Parking Ticket Capital of the World." And Hermosa Beach's mayor worries that the April ballot measure might jeopardize state approval for onshore drainage of heavy oil from state tidelands.

Am I missing something here?

The frozen South

Warren R. True **January 20, 1986**

Pipeline construction has fallen on hard times. That assessment will surprise no one.

This year's report on OGJ's annual pipeline construction survey projects a decline of almost 12% in construction for 1986 and beyond.

Some pockets of activity, such as the U.S. West, the North Sea, and the Middle East, continue. But pipeline construction activity generally has tracked the decline in petroleum prices.

The annual meeting of the U.S. Pipeline Contractors Association, therefore, could be expected to be subdued. But there may be another reason.

Gathering each year at the staid, old (and expensive!) Boca Raton (Fla.) Hotel & Club, contractors take a breather from the cold, snow, sleet, and mud of their construction spreads and sites to the north. After morning meetings, golf and tennis are priorities; business is discussed in the clubhouses or beside the pool.

Attendees at the 1985 meeting arrived on Saturday and Sunday, shedding coats and sweaters as they landed in the South Florida sun. Ahead, they knew, lay 3 or 4 days of what the rest of the country was only dreaming of as it shivered in the grip of the latest and most severe winter storm.

But that storm pursued the contractors to Florida.

They awoke Monday morning to temperatures in the low 30's. A stiff northerly wind set golf course flags stiffly southward and bowed tennis nets. Courses and courts were empty.

The tall hallways of the hotel, built to capture cooling sea breezes and channel warm air to the ceilings, became wind tunnels. Hunchbacked guests, layered in sweaters and coats, huddled together to hustle between restaurant and room, bar and ballroom in vain attempts to escape the cold.

A year's perspective on that frigid meeting has shown its symbolic significance. Optimistic projections for 1985 have proven hollow: Pipeline construction, which in 1984 had shown strong signs of recovery, began to drop sharply — so much so that in the final quarter of 1985, applications to the U.S. Federal Energy Regulatory Commission (FERC) for natural gas pipeline projects dwindled to a trickle.

U.S. pipeline contractors return to Boca Raton next week remembering last January's blast and 1985's chilling business activity. Hopes for warmer times will abound.

The rubber ruler

Gene T. Kinney **February 24, 1986**

In publishing, as in the oil business, it is necessary to have devices to make things come out even.

Journalism has been defined as the ability to meet the challenge of filling the space. Despite the careful instructions of editors, writers don't always meet that challenge with precision. They either run out of ideas before reaching the bottom of a column on drilling, or their words spill over to a page that is already allotted to enhanced recovery.

Hence, the necessity for a story that can be stretched or compressed a few lines without shortchanging the reader. In publishing, this very handy gadget is called a rubber ruler. No editor could get to press without one, and there's no harm done.

Capital spending is the rubber ruler of the oil business. But the consequences of stretching and compressing are vastly different.

In 1981, when U.S. investors still thought the sky was the limit, the ruler was extended to incredible lengths. Oil and gas men couldn't spend money as fast as it was being thrown at them. The beginnings of the crash were already visible. But before the cash machine could be turned off the next year, industry poured a record 34% of net wellhead revenues back into the ground.

Since then, oil companies have reversed course. In order to boost present returns to stockholders and survive, they have put money almost everywhere but into drilling. To make the balance sheet come out right, capital spending has been cut and cut again.

This year, the rubber ruler is snapping back to the shortest length on record. Confronted with the prospect of further erosion in prices, industry had already made plans to trim budgets again. Based on an estimated 10%

decline in prices, Oil & Gas Journal had forecast a 12% decline in drilling expenditures. At $18.5 billion, projected drilling outlays yielded an apparent reinvestment ratio of 17%, the lowest in recorded history.

Then Saudi Arabia brandished its new version of the oil weapon, first used to create a shortage and a price rise in 1973 but now deployed for opposite reasons. So oil companies are back at the drawing board. Once again, the rubber ruler is out. This time, in order to pay bankers and stockholders out of sharply reduced revenues, managements are generally cutting capital spending 20-33%. Some cuts are greater than that — whatever it takes to salvage a decent bottom line today, although with great consequences for the future.

Oil companies are on the horns of a dilemma. If $15 oil is the shape of the future — the Saudis would obviously like to convey that impression — investments that promise a good return only at $20 or $30 don't make any sense, even to those with the cash and credit to afford them. But what if companies gear investments to that assumption, as they appear to be doing? Then their reserves and production will soon go into decline and OPEC production and prices will rise accordingly.

Thus, the Saudi strategy seems to be working to perfection.

Son of Eniac

Leo R. Aalund **March 3, 1986**

This is the third year Oil & Gas Journal has published a special report on microcomputers.

This year we decided the time was ripe to take a look at what the entire oil industry was doing with microcomputers and what users wanted to be able to do with them. We thought by now enough experience and maturity had been acquired with this technology to reveal some useful applications and solid trends.

Microcomputers have, of course, received more publicity than practically any new arrival, technological or otherwise, on the American scene.

Was it deserved? Most of it was.

Entire disciplines are now practically dependent on the microcomputer.

In others it is the key to efficiency or a very handy tool.

OGJ has been keeping up with computers and the oil industry for many years. Starting in 1958, we launched a pioneering annual report that detailed computer control projects and installations. We went from frontier technology and a few conspicuous projects then to applications so numerous that we had to quit counting in the mid-1970s.

In late 1977 we published a landmark address by Dr. Theodore J. Williams, then president of the American Federation of Information Processing Societies. His topic was computer control in the 1980s.

He very correctly assessed the existing situation -- and the future -- by saying, "The microprocessor has taken the computer field by storm."

He went on to say, "There is no question that many wonderful things are going to be done to us -- as well as for us -- by and in the name of microprocessors in the relatively near future."

With due credit to Dr. Williams' excellent insight, we wonder if even he has been surprised.

From a broader perspective, anyone has to be surprised at what has happened in our brief computer age. The complete history of petroleum applications has been reported, week by week, as it unfolded, by editors and technical writers still on the OGJ staff who duly recorded the introduction and progress of computers in the industry.

Only 40 years ago last month the electronic numeral integrator and computer (Eniac), foreunner of our modern computers, was officially introduced at the University of Pennsylvania.

It employed vacuum tubes, weighed 30 tons, and required a hefty air conditioning system of its own. Its computing power can now be more than matched by a tiny silicon chip.

No business surpasses petroleum in exploiting the possibilities opened up by this incredible feat of technology. And, with the special report in this issue, OGJ continues its commitment made a generation ago to serve as the leading chronicler of how computers can be used for profit in all phases of petroleum operations.

20 cents/bbl

John L. Kennedy **March 17, 1986**

Don't panic. That's not the latest price of Brent blend for May delivery.

But we felt that what Journal readers need most now is some good news about the price of crude.

Since I can't bring you any, I'll do the next best thing: I'll try to convince you it has been worse.

Probably not during my lifetime or during the career of anyone currently younger than normal retirement age.

But never mind that.

Recently, Max D. Eliason, executive vice-president of Skyline Oil Co., Denver, sent a copy of Humble Oil & Refining Co.'s Price Bulletin No. 162, dated May 13, 1933. It gave notice that Humble would pay, for crude from several West Texas counties, a "posted per barrel price for crude oil purchased and delivered for its account into the lines of Humble Pipe Line Company . . ." of $0.20.

That's 20 cents a barrel.

It also noted that "Determination of quantity, deductions, adjustments for temperature, etc. . . ." for oil purchased in Texas would be in accordance with the rules and regulations of the Texas Railroad Commission.

Deductions? Adjustments for temperature?

At 20 cents/bbl?

Talk about driving a hard bargain!

But crude had been even cheaper — 10-15 cents/bbl. The Great Depression, coupled with another "disaster" — discovery of the East Texas field — were responsible for many of the industry's market problems in the early 1930s.

In the summer of 1933, when Humble posted 20 cents/bbl for several West Texas crudes, the tank car price of regular gasoline was 2 5/8 cents/gal.

Even though crude and products were virtually being given away, not everyone in the oil business was moving without leaving a forwarding address.

Technology was being developed and proven in drilling, completion, production, processing, and refining. Examples include new bit designs, drilling data recording devices, gun perforating, hydraulic pumping, polymerization, propane deasphalting, desalting.

This and other technology and equipment has served the industry ever since, and led to ever greater efficiencies and capabilities.

And new fields were being discovered, even though East Texas temporarily swamped the crude market.

Oh, I know. The 1986 equivalent of 20 cents/bbl in 1933 is, well, a lot more.

Maybe way up over $2/bbl, based on the producers price index.

But it isn't even anywhere near this week's minimum posting for West Texas Intermediate.

Feel better now?

Origins

Richard A. Corbett **March 24, 1986**

The week before deadline for Oil & Gas Journal's annual refining report usually finds the refining editor attending the National Petroleum Refiners

Association refinery and petrochemical plant maintenance conference.

This year's meeting was in San Antonio, home of the historic Alamo, particularly revered in this 150th year of Texas' independence from Mexico.

In addition to coverage of this important meeting, the finishing touches to this year's refining report had to be completed by the following Monday. This meant several hours in a hotel room crunching survey data and polishing the format of contributed articles.

And those hours had to be weaved into a busy schedule of informative technical presentations, interesting exhibits, and inviting hospitality suites.

Alamo Plaza was in full view from the hotel window as work proceeded on the special report. Thoughts of the 13-day struggle that began there Feb. 22, 150 years ago triggered curiosity about the origins of OGJ's annual refining report.

The origins are obscure because it is difficult to find clear references to a regular refining report before 1926, after which it apparently became an annual springtime event called, for many years thereafter, the Refining Number.

But an early major work bylined by H. G. James, OGJ's refining editor back then, appeared as a supplement to the Journal on May 22, 1919. A wealth of information was provided by the report, with statistics ranging from the investment in each refinery to the number of tank cars and miles of pipeline owned by each refinery.

There were about 400 operating refineries in 1919 that processed about 1.4 million b/d of crude oil. In those days, crude runs were the important number, not crude charging capacity, as today's survey reports.

Contrasting operating practices were evident back then as this quote from the report reveals: "Outside the large cities, most eastern refineries are of small capacity. The owners take their profits and chuck 'em down in their jeans and enjoy life, while the western refiner takes more than his profits and enlarges his plant — and sweats blood."

As much as the survey has changed since 1919, several refineries still exist today, although some names have changed: The Texas Co., Port Neches, Tex., Standard Oil Co., Richmond, Calif., Vacuum Oil Co., Paulsboro, N.J., and Sun Oil Co., Toledo, Ohio, to name a few.

During the trip back to Houston, thoughts of how refining has grown to the shape of today's industry were accompanied by thoughts of Sam Houston's trek eastward to that historic battle that would secure Texas' future as a republic and a state.

But those historical musings didn't hide the fact that the refining report was still due Monday.

And like Houston, spurred by a renowned battle cry, my trip was similarly motivated by "Remember the deadline."

Breaking in weevils

Michael Obel **March 31, 1986**

Sometimes the hardest part of a veteran rig hand's job is dealing with new guys.

News of Elf Gabon's successes off West Africa reminds one such rig hand of the challenges he faced while drilling a wildcat off Gabon.

Nicholas Edeleanu, an Englishman who worked for Keydril Co. as lead floor hand on the Key Gibraltar jack up a few years ago, says one of the most difficult problems for new Gabonese roustabouts was the industry's safety code.

"Many of them had not lived in towns before, so equipping them for work offshore was quite a task," Edeleanu recalls. "They liked the hard hats and found what seemed to be hundreds of ways to wear them. But they had trouble at the other end of their bodies."

Many of the Gabonese workers could not get used to their safety boots and so removed them. Most complied with safety rules after being sternly ordered to put their boots back on.

That appeared to be the end of the matter until a few days later when Edeleanu noticed something strange about the boots on a couple of new workers.

He checked and found that the men had cut off the boot soles, then laced the tops over their feet and ankles to make it look as if they were complying with the rules.

There also is a painfully true story of the classics student from the University of Missouri who in the summer of 1977 got a job as a roustabout on a Cactus International rig about 90 miles off Morgan City, La.

When the crane operator lifted that man off the crew boat, everything about him shouted "weevil."

It was a long summer.

For the longest time the new hand couldn't find the key to the V-door. "Is it in this pile of stuff?" he asked, holding an armful of angle iron he had collected from all over the rig.

Eventually the weevil gained enough confidence to offer new insights into the oil industry.

"You know why a cheater bar works?" he once asked the driller.

Misreading the driller's silent stare, the roustabout went on. "It has to do with what's called the Law of Moments. A Greek mathematician who died about 3,000 years ago discovered it. Isn't that interesting?"

Something about the way that driller kept staring at him made the new man realize it might be best to leave the connection between cheater bars and Greek mathematicians alone for awhile.

Lucky for the contract drilling industry, that youthful Yankee decided on white collar work. As things turned out, he went into journalism and eventually got a job with Oil & Gas Journal, where he can recall in print a long summer with the patient people of Cactus International.

Gloom and doom

Rick Hagar **May 12, 1986**

The young loan officer was miffed at a newspaper headline that said growing problems with energy loans will cause more bank failures.

"I just don't believe that we or other prudent banks are in trouble due to our energy loans," he complained to his department manager.

"Son," said the manager, "you weren't around back when most of these loans were made. You need a little education."

The manager pulled from his bookcase a yellow and black magazine that was as thick as many city telephone directories.

"This is the Sept. 27, 1982, issue of Oil & Gas Journal," the manager said. "Read this on page 118."

The story detailed the results of a 1982 survey of bank energy loan departments in Houston, Dallas, Chicago, and New York.

It said bankers, who made loans in 1980-81 assuming that skyrocketing oil and gas prices would continue without pause, were beginning to believe that prices had cooled. But they also believed prices would again take off in 1984.

After 1984, the bankers assumed in 1982, prices would rise 6%/year, which many considered moderate compared with the 1980-81 assumption of a 10%/year escalation rate.

The article said nominal base prices used by bankers in assessing the value of their 1982 loans were $30-32/bbl for oil and $6/Mcf for deep gas, compared with $36/bbl and $8-9/Mcf for 1980-81 loans.

Furthermore, maximum price assumptions in 1982 were $75/bbl for oil and $12/Mcf for gas, down from the previous expectation of $100/bbl for oil but unchanged for gas.

"How could industry have let us use such loan value criteria?" asked the incredulous loan officer.

His manager offered a sad smile and said, "At the time, they thought we were being unjustly harsh, overreacting to a temporary glitch in the rise in prices."

He explained how the petroleum industry greatly expanded to keep up with the expectations of the early 1980s and was now, due to lower prices, dismantling the buildup.

"But what happened?" asked the loan officer. "How could we have been so wrong with our expectations?"

The manager returned to the 1982 OGJ and pointed to a headline on the story that followed the loan criteria report. It read: "Yamani revives threat to cut market crude price."

"The 1982 loan value criteria seem silly in 1986," said the manager.

"But before you start spouting off about how stupid we were, just remember that today's talk of gloom and doom probably will seem just as silly in another 4 years."

The vulture fund

John L. Kennedy **September 8, 1986**

All the way down the back slope of the drilling boom that peaked in 1981, industry spokesmen have been using vague terms to describe what's happening in the industry and what they are doing about it.

The latest crop of corporate annual reports, a key source of information for the OGJ 400 report in this issue, turned up some examples.

Company chairmen told shareholders through their annual reports that after "adjustments" in 1985, 1986 was expected to be "a difficult year."

For some, that expectation promised to be wildly optimistic.

When West Texas intermediate fell to $18/bbl, the modern, forward-looking company began "positioning itself for the upturn."

But at $10/bbl, some had to abandon that strategy and position themselves for next week's meeting with the bank.

And there were those company leaders that were "looking over the valley." I'm not sure which valley that was. Most of the petroleum industry has been looking over the Grand Canyon.

"Unsettled markets" is another problem faced recently by oil and gas companies. Describing the crude oil market during the past 6 months as unsettled is like describing today's drilling rig fleet as "underutilized."

Maybe those companies were in another market.

Of course, while looking over the valley during a difficult year of unsettled markets, part of the process of positioning oneself for the upturn is to take advantage of "opportunities for building a strong base."

That's one way to make a virtue of necessity while counseling patience to shareholders. But as a strategy, it's not altogether clear.

In fact, I'm a little unsure what many of these phrases mean.

But one company has taken action that may mark the end of the trend to describe the industry's plight in euphemisms that put the best face on a bad situation.

Harrington & Co., a New York investment banking firm, announced last month the closing of its "oil and gas vulture fund."

Mark Harrington, president, said opportunities targeted by the fund would include exploration prospects acquired via farmout from "financially distressed operators," prospects abandoned by operators "for lack of available capital," and acquisition of producing properties from "financially squeezed" operators.

This first oil and gas vulture fund will be used "to exploit a number of investment opportunities" and may even make a future edition of the OGJ 400.

A vulture exploiting distressed, financially squeezed companies that lack capital.

Now that I understand.

Milestones

Bob Williams **September 22, 1986**

The oil and gas industry loves numerical milestones.

Just as McDonald's tracks its latest billionth burger, oil and gas operators revel in operating watersheds. You may not get excited about another record drilling time for a North Slope rig, but it tells something about the push for efficiencies in that harsh, costly desolation.

This column celebrates a milestone that could not be more timely or significant: the tanker loading of the 5 billionth bbl of oil from Prudhoe Bay oil field on Alaska's North Slope.

That historic barrel of Sadlerochit crude entered the Trans-Alaska Pipeline System (TAPS) at about 4:30 p.m. Sept. 8. It took a 5 1/2 day journey traveling about 6 mph for 800 miles, arriving at Valdez terminal at 5:28 a.m. Sept. 14. It was loaded into the Admiralty Bay tanker and shipped to a West Coast refinery.

Why is this important? Because it represents the halfway point of economic depletion for a field that has accounted for about 10% of U.S. oil supply since 1977. Prudhoe Bay started up June 18, 1977, with initial output of 300,000 b/d and built quickly to 1.5 million b/d.

It's also the midpoint for the centerpiece of a province that could continue to contribute that much to future U.S. energy security. And it's the latest notable number for a project that proved industry's ability to preserve a fragile arctic ecosystem while it extracts vital oil supplies.

How about some other numbers? Ten billion bbl, Sept 1, 1986. These represent the estimated oil potential of the Arctic National Wildlife Refuge (ANWR) and the original date (actually Sept. 2, given the Labor Day holiday) on which U.S. Fish & Wildlife Service (FWS) was to deliver a report to

Congress on hydrocarbon potential and wildlife in the refuge. If there's another Prudhoe Bay in the U.S., ANWR's the best bet.

Save for a rank wildcat reportedly yielding good oil shows on native lands, no drilling has occurred in ANWR. Development is forbidden throughout the refuge and exploration allowed only on native lands there. Congress may — but isn't obliged to — render a decision on whether to open ANWR to possible oil development. That's moot now because Trustees for Alaska sued to block the FWS report. Many see ANWR oil development as Ecology Issue No. 1.

Here are some more milestones, including potential ones:

• 2000 — A year that will come and go before oil is produced from ANWR, says ARCO Alaska Inc. Pres. Harold Heinze.

• 1988-90 — When Prudhoe Bay is expected to begin a sharp decline.

• July 14, 1986, and 663 — The date and tally for the historic low of the Hughes Tool Co. count of active rotary rigs in the U.S.

• 6.962 million b/d and 43.3% vs. 29.3% — The 4-week average level of U.S. crude and products imports as of Sept. 5, and a comparison of the U.S. import dependence level with the same period a year ago.

• Sept. 1, 1986 — OPEC reintroduced a production ceiling in a step to regain control of world oil markets.

Oilmen know October 1973 as another milestone — the beginning of the Arab oil embargo — that spurred the U.S. energy crisis, thus rescuing Prudhoe Bay from an environmental quagmire and triggering congressional approval to build TAPS.

What date will history judge as the milestone for the next energy crisis?

The 'Dallas' view

Rick Hagar **October 20, 1986**

As visitors to Houston were being driven to Galveston Island for a day at the beach, they noticed a large sign identifying the Texas City exit from the freeway.

The visitors asked the driver to take that exit because they wanted to see Clayton Farlow's refinery.

Clayton Farlow is the husband of Miss Ellie, the grand dame of the television series "Dallas" and the mother of bad guy J. R. Ewing.

It's shocking to realize that many people, even those who live in producing states, believe that characters and situations depicted in "Dallas" are true to life.

In the show, Clayton has decided to sell the refinery in order to start a horse breeding business with Ray Krebbs, Miss Ellie's ranch foreman and the illegitimate son of her first husband.

A reasonable person might ask, "So what?" Well, industry should be aware of some of the impressions this year's show is leaving with viewers.

Just like many U.S. independent producers, J. R.'s company has seen its revenues slashed by the oil price collapse. And, similar to actions taken by other producers, J. R. is shutting in marginal production.

But unlike real producers, J. R. is developing a plan to make sure that world oil prices rise. He apparently is putting together a commando team to raid Saudi Arabian oil fields. And the show implies that most Texas oilmen would be in sympathy with such a plan! These TV producers view the Saudis as religious fanatics bent on destroying the Texas oil and gas industry.

Meanwhile, Cliff Barnes, J. R.'s arch enemy, also is scrambling to help his producing company. It's his wife, however, who comes up with the idea that gives the best hope of saving the day — an oil import tariff.

Cliff, of course, steals credit for the idea from his wife, who happens to be J. R.'s cousin, and immediately drums up widespread support. But in a short-lived power struggle, J. R. grabs control of the oil import fee committee from Cliff and gives it to Donna Krebbs, the estranged wife of J. R.'s half brother and the widow of a powerful national politician from Texas.

In the television show, oilmen want the import tariff only because they are running out of cash to support their extravagant lifestyles.

Again you ask, "So what?"

Well, maybe something extra should be done to educate those who are anxious to see Clayton Farlow's Texas City refinery.

It's frightening to think that their "Dallas" view of the industry may someday dictate the actions of their congressmen on oil matters.

Booms and scars

Jim West **October 27, 1986**

Every year about this time, Oil & Gas Journal takes a close look at how U.S. independent producers are faring.

We've been doing that since 1977, recording the ups and downs of more than 200 independents in some years.

This year, our coverage of earnings and revenues for the first 6 months of the year has shrunk to a list of 170 operators. Only about one-third of them showed a profit in the first half.

Many producers that regularly appeared in our yearly report on independent operations have fallen prey to hard times, characterized by things like bankruptcies, in the Crash of '86.

As a result, the industry boom that went bust has shown up in a different form and in a different arena. Where else but in Texas?

The National Law Journal reported last month that in Austin, while lawmakers were debating a state budget crisis, bankruptcy lawyers were reliving the Thursday Massacre.

Lawyers were recalling Aug. 28, when R. Glen Ayers, federal bankruptcy judge for the western district of Texas, opened a 105 page docket with three to five proceedings on every page. It took the judge 2 days to complete the session. The docket call filled Austin's largest federal courtroom wall to wall with lawyers, forcing court personnel to deny admission to everyone else.

The court's western district covers a sprawling area, with two judges shuttling about 340 miles from Waco to Del Rio and about 575 miles between San Antonio, Austin, and El Paso. In the district is the Midland-Odessa area, the center for oil and gas operations in the Permian basin.

The number of bankruptcy filings in the western district is about 90% greater than last year, the National Law Journal reported. The bankruptcy boom has clogged the courts and created delays in handling cases under Chapter 11 of the bankruptcy code.

Although the volume of cases is large, the western district has not seen any monster filings like the one in New Orleans by the H. L. Hunt interests of Dallas. That one involves protection from bank foreclosures on more than $1.5 billion in loans.

No doubt about it, the oil price collapse has inflicted permanent scars on the industry.

For example, managers won't forget the time they handed the last pay checks to employees caught in a layoff. Neither will employees, past and present.

"The hardship visited upon these fine people and their families, although inevitable in this environment, is the worst part of this time of troubles," says Hugh J. Kelly, president of Ocean Drilling & Exploration Co., New Orleans.

Drilling contractors will remember their sinking feeling as they watched perfectly good equipment being written down to salvage value. And bankers will have an indelible picture of a $10 price tag on a barrel of oil as the "worst case" measure of the economics of a loan sought on reserves.

As with most scars, these will never completely fade away.

Here we go again

Leo R. Aalund **November 17, 1986**

What do these incidents and messages tell you?

• A door-to-door peddler selling front yard gas lights.

• The light company informing you that you will qualify for a lower electric rate if you have an "all electric home."

• An oil company service station giving away drinking mugs, balloons, or bubble gum with each fillup.

• A network news anchorman asking on camera with a tone of concern an oil expert from Boston if the latest OPEC production control agreement will "drive the price of our gasoline up again."

• A car salesman who doesn't mention gasoline mileage once during his sales pitch.

If you guessed that you were in a time warp back to the early 1970s or that you were living in 1986 in the U.S., you would be correct in either case.

Such incidents are striking, bringing a strong sense of deja vu. We are back to coping with cheap energy, a result of the plunge in oil prices early this year. This manifested itself in dramatic bank failures, bankruptcies, and big layoffs in the oil industry. Now cheap energy is evident in little deals, transactions at the most basic levels of personal business.

The light company, for example, sees its customers switching to cheaper gas. Two appliances it dearly wants to keep on the bill are the electric cook stove and electric hot water heater.

The gas company, however, isn't standing still. A bright idea there is to revive or relight the gas light and get some income from the new installation and the subsequent glow. When gas is cheap, this is only pennies a day for the consumer, even though most of the lamps burn day and night. But several thousand turned on or installed add up to a nice piece of steady change for the gas company.

Many gas lights have stood cold since the late 1970s. After all, even the most naive consumer would have balked about a day and night gas light when the air conditioning bill was running more than $500/month and gasoline was sky high. The urge to conserve was a bit stronger then.

Gasoline is now so cheap that many drivers are zipping past the "pay first" pumpers straight to real eight bay stations — any station — charging perhaps 5¢/gal more. Dealers may even start wiping windshields and giving away road maps again to get attention.

It is too soon to say that public energy behavior will revert to the conspicuous consumption of the good old days. But there are signs, unless OPEC steps in, it will. The oil industry should get ready for this.

Let's talk about 1987

John L. Kennedy **December 15, 1986**

Nineteen eighty-six has been a year like no other. It cries out for some sort of special recognition. So we're doing something unusual.

We're publishing only four issues of Oil & Gas Journal in December.

It's our way of giving coming generations of oilmen a benchmark year from which they can measure the events of their times.

When a future planner is cautioned, "Don't forget what happened in 1986," the likely reply will be:

"How could I forget? That was the year the Oil & Gas Journal published only four issues in December instead of five!"

Well, it's possible.

What is certain is that our next issue, the final issue of the year, will contain our Worldwide Report. It's full of statistics and analysis on worldwide production, reserves, and refining.

The hard numbers in this special report tell what the net effect of all the agony, the failures, and the successes was on the industry last year.

Of course, that issue will contain all the significant industry news that occurred during the previous 2 weeks. And it will have the technology and the statistics found in every issue.

What we're really doing is combining the last two issues of the year.

In addition to its value as a memorial gesture appropriate to the passing of a year we hope is never equalled, this combination of 2 weeks' news in one issue will, I admit, slightly reduce our costs for the year.

After all, there is no stigma attached to combining operations these days. Many companies are thankful they have any operations to combine.

We're only combining the news of the last 2 weeks in December into one issue.

For us, it's a big deal. But considering what has happened to many in the industry during the past year, we consider ourselves fortunate.

And this minor bit of belt tightening, when the Christmas holidays tend to replace business as usual, will help preserve the strength of our editorial service in the future.

But enough of past troubles. We'll give 1986 what it deserves as one of the most significant years in the history of the petroleum industry: a Monday of silence.

No eulogy. No more hand wringing.

And no Oil & Gas Journal on Dec. 22.

Now let's talk about 1987.

Yamani's predictions

Rick Hagar **January 5, 1987**

Are you aware that Ahmed Zaki Yamani, Saudi Arabia's former oil minister, predicted in 1969 there would eventually be a dramatic collapse in world oil prices?

The prediction came in a speech Yamani gave at an American University seminar in Beirut (OGJ, July 14, 1969, p. 53). His argument was that major oil companies, deprived of their Middle East producing properties through nationalization, would lose interest in maintaining wellhead prices at high levels.

Elimination of this profit margin upstream would force the majors to shift their profit focus downstream to refining/marketing operations. Consequently, he said, majors would put their full weight behind efforts to drive down crude oil prices.

Producing countries, meanwhile, would find themselves in direct competition as producers and sellers, he said. This competition also would depress prices.

His conclusion was that such a scenario could have only one outcome: a dramatic collapse of the price structure with each producing country trying to maintain its budgeted income in the face of declining prices by moving larger volumes of cheaper and cheaper oil into an already glutted market.

Although a lot happened in the meantime, Yamani's scenario sounds very similar to what eventually developed. Oil & Gas Journal readers are painfully aware of the slow slide in world wellhead prices after 1981, then the dramatic collapse during last year's first quarter.

Yamani predicted that majors would reduce or even eliminate exploration and production operations to become mere lifters of oil. Though it has other causes, too, last year's rig count slump may reflect that trend.

Yamani also predicted that producing countries would begin to participate more in downstream operations, previously the stronghold of majors only. As OGJ readers know, this outlook is a growing reality.

And Yamani said producing countries would eventually form a cartel in order to bring some stability to world prices.

He wasn't speaking of OPEC. The organization was 9 years old at the time of his American University speech. More than likely his cartel is represented by the new bond between Iran and Saudi Arabia, the two countries that joined forces to push through the latest OPEC production agreement coupled with a return to posted prices.

Considering how accurate Yamani seemed to be with his 1969 predictions, it's important to note one more: Unless major oil companies and producing countries strike some balance in upstream and downstream joint participation, he said, there will not be a return to stable prices.

"Neither we nor the majors would be powerful enough to maintain stable prices," he said. "So what we are proposing is that, via participation upstream and downstream, we should get together and grow within a stable setup in which the parties both have the desire and ability to maintain prices."

It's something to keep in mind as 1987 gets under way.

North Slope Club Med?

Bob Williams **February 2, 1987**

In the quest to deliver the best information that readers of Oil & Gas Journal can use in their jobs, OGJ writers/editors occasionally must resort to the dreaded analogy to make a point.

Usually, they are to be avoided. It's best to say a thing directly. For example, the news editor promises a Fate Worse Than Death for the writer who lets the next description of a rig or plant as "the size of X football fields" slip past.

Regular readers of this space will recall the undersigned's frequent analogies, in order to make a point about the industry, to hiking/climbing/backpacking misadventures in a previous incarnation as OGJ West Coast Editor and resident of the People's Republic of Southern California.

Many West Coast Journally Speaking columns have dwelled on the logistical hardships and high costs encountered in scouring Alaska's wilderness, to underscore industry's own problems in that harsh yet sensitive environment. Perhaps, some hoped, the change of assignment to Tulsa might have put an end to such daisy-sniffing ramblings. Not so.

My old friend, Chevron Corp. Vice Pres. Clair Ghylin, has delivered a beaut of an analogy to make a point about the U.S. government locking up the most attractive petroleum prospect in North America, the Arctic National Wildlife Refuge's coastal plain.

Clair put a price tag to locking up ANWR as a pristine wilderness experience. Taking U.S. Fish & Wildlife Service's calculation that there is a 5% probability the ANWR coastal plain oil resource is about 30 billion bbl and assuming a world oil price of $20/bbl, he calculates a cost of $600 billion in replacement oil imports. That's the tab for banning the drillbit from about 1.5 million acres of additional wilderness in a state where the wilderness area already exceeds the area of the state of Washington.

Although the region has had only about 200 visitors/year since ANWR leasing and development became an issue some years ago, Clair generously assumed 1,000 visitors/year and arrived at a $20 million subsidy per visitor.

"That's a pretty elitist backpacking experience," he said.

Oops. Exposed. I confess to having been one of those elitist North Slope backpackers, arrogantly indifferent to the costs of enjoying a rare wilderness experience.

Yes, I've been to the Prudhoe Bay National Forest, the tallest nonmetal sight for thousands of square miles north of the Brooks Range.

It's about 12 ft tall, made of plywood in the shape of a fir tree, and it's evergreen so long as the green paint can withstand -50 degrees F. and 70 mph winds in February.

Slump and bump

Michael Obel **February 9, 1987**

Stepping into the control room of Amoco Oil Co.'s huge resid hydroprocessing unit (RHU) at its 400,000 b/cd Texas City, Tex., refinery, the first-time visitor finds himself in what could pass for the control room of a battle cruiser out of "Star Wars."

The lighting is dim, the noise level low, and numerous Technicolor terminals display whatever operational data are desired. One can almost hear Darth Vader inhaling and exhaling in his lordship's diabolically raspy way. The only thing missing is a crew of intimidated workers whispering in a highly esoteric nomenclature.

The spell starts to wear off, though, as soon as one begins visiting the workers. The unit is run by a crew of busy but relaxed operators, none of whom appears to be awed by the exceedingly complex electronics that allows them to take resid and upgrade 60% of it into more valuable products.

Consider, for example, the way employees talk about their jobs. One operator, asked how he might fine tune the temperature in a reactor, explained that he might "slump and bump" the catalyst bed.

This isn't the language of chemistry, engineering, or electronics. It's everyday language like one can hear anywhere in South Texas.

Slump and bump refers to the process of slowing a reactor's ebullating pump to lower the level of the catalyst bed in the vessel. Lowering the catalyst level reduces the amount of activity in the reactor, thus cutting temperatures. That's important because if the temperature inside an RHU reactor gets too hot, thermal cracking might begin and coke might be formed. The idea is to upgrade resid, not crack it into coke.

Once temperatures come back within an acceptable range, the operator simply increases pump speed, thus bumping up the level of catalyst. Activity increases. So do temperatures. Things are back to normal.

From the worker's or visitor's perspective it's simple, just a matter of requesting data, inspecting the displayed numbers on a color screen, and reducing a specific pump's speed. But that simplicity — exemplified by the operators' down-home phraseology — is possible only because of sophisticated electronics.

The longer a visitor stays, the more engaging the RHU's control room becomes. He begins to think that with a little instruction and practice at a terminal, he too could shepherd a few thousand barrels of resid through one of the reactors at Amoco's $600 million RHU.

Leaving the control room, one gets the impression that high tech may spawn the occasional Darth Vader. But at its best, it is more likely to produce a relaxed atmosphere where one worker can casually lean toward another and say, "Hey take a look at No. 3. Might want to slump and bump."

The PLCA returns

Warren R. True **February 16, 1987**

Attendees will remember the 1986 U.S. Pipe Line Contractors Association annual meeting in Boca Raton for reasons far darker than their having seen the second year of record-breaking Florida cold.

Why the PLCA holds its winter annual convention there needs no explanation. Many of the people who spend much of their professional lives digging ditches in sand, mud, ice, and snow prefer a place away from those elements to compare business notes.

In 1985, however, the wind and cold followed them to Florida, a disconcerting development surely not to be repeated in 1986.

Yet, Monday afternoon of the 1986 meeting, record-breaking cold again blew across the golf course and tennis courts and turned the old hotel's vaulted halls into chilly wind tunnels.

By Tuesday morning, ice had formed on the Challenger 7 space shuttle's launch pad, barely 2 hr north of Boca Raton. The nearness of the catastrophe that befell Challenger 7 added to the shock we shared with the nation. Our cold snap was the same that had chilled that launch pad and the shuttle.

By Tuesday evening's convention banquet we had all seen the pictures: debris falling and twisted columns of smoke spiraling outward, hanging interminably in the clear, cold blue.

Table comments were couched in cliches that help us deal with such events by shielding our full comprehension. Yet the monotone still seemed to echo in the dining hall: "We have no downlink . . . obviously a major malfunction."

By Wednesday morning, flags around the hotel compound and along the coastal highway to the airport flapped at half staff in the diminishing, warmer breezes.

Last month marked the first anniversary of the Challenger 7 explosion and the 1986 PLCA convention. PLCA, seeking to evade a third year of record Florida cold, returns to Boca Raton this week.

Awaiting them will be another Oil & Gas Journal Pipeline Construction Report. Its sober figures reflect what the operating industry has told us about its worldwide pipeline construction plans.

But the weather promised to be warmer this week. And pipeline builders around the world do have some challenging work ahead that also will help lift the spirits — and revenues — of some.

Some readers at the meeting will pause to look northward and remember last year.

Computer safari

John L. Kennedy **March 2, 1987**

"Seven squirrels at mile post 213.20, one at 213.50, one at 214.00, and one at 214.01."

Oil and gas companies increasingly need this type of information. Not what brought seven squirrels together, but the distribution and population of all kinds of wildlife, especially those that are rare or endangered.

When just one state contains hundreds of different species in thousands of locations, it's no surprise that the information tool of the 1980s — the electronic database — is called into service.

Dwight Holing of Woodward-Clyde Consultants, San Francisco, describes the practical use of computerized inventories of rare and endangered wildlife in an article he has prepared titled, "Keeping track of the birds and bees."

According to Holing, 38 U.S. states operate their own "natural diversity" databases. They can be used to help prepare wildlife and other reports that must accompany project applications. For most reports, a field study is still required. But the database can quickly prepare the way for a field study, says Holing.

California's natural diversity database, maintained by the Department of Fish and Game's Natural Heritage Section, is the largest, says Holing. It contains more than 16,000 locational records for more than 1,200 species.

California has 5,000 native plants and close to 1,000 native vertebrates. One third of the plants and one half of the animals occur naturally nowhere else on earth. Those are the most likely candidates for state and federal endangered species lists.

Information in the database is compiled from published reports and observations by a contributing network of field biologists and naturalists. A staff of 25 operates the system.

The system can supply map overlays and reports. Data are retrieved by specifying the geographic area, either by quadrangle, county, or latitude/longitude; a species or natural community; or a status (designated by state or federal as endangered, threatened, or rare).

San Joaquin antelope squirrels, according to the database, were seen in 1981 at the milepost locations indicated along a proposed pipeline related to a cogeneration plant. The species is listed as threatened under the California Endangered Species Act and is a candidate for protection under federal law, says Holing.

Woodward Clyde was hired to do a detailed wildlife and vegetation report on the project.

A database of information on wildlife used by oil and gas companies to study a project area is yet another example of the industry's resource development to coexist. Like a subsurface safety valve. Or an arctic pipeline monitoring system.

The list is very long.

The T word

Bob Tippee **March 16, 1987**

It has swept through the U.S. oil industry like a hurricane through the Gulf of Mexico. People pay money for it. Some might even give up barbecue for it.

You know what it is. The T word. Tariff.

More particularly, it is the tariff (or import fee or floor price or whatever) controversy because the issue hasn't advanced beyond that stage. Do you support it, or do you not? In some sections of oil country, one answer gets you in the door, while the other might well get you a boot where a saddle is supposed to go.

The tariff inspires strong feelings — in favor if you're involved in the domestic drilling/producing industry, against in the case of refiners and petrochemical plant operators, a few large producers, and suppliers of foreign oil to the U.S. market.

This column will not argue the merits of the issue. It simply will make some observations about the passions surrounding it.

One observation concerns the source of those passions. Some say desperation has driven hard-pressed oil producers to break character and ask the government to intervene in the market on their behalf. Times are tough. Producers and their suppliers are going broke in herds.

But it must go beyond desperation all the way to frustration.

Producers sense, probably correctly, that voters and power brokers outside oil country either don't know enough to care or don't care enough to know about what's making them desperate.

The drilling slump inevitably has become the domestic production slump, which inevitably must become the next oil supply crisis as imports surge in the U.S. oil supply mix. Yet all producers hear from outside oil country is whining about the price of gasoline.

No wonder they're frustrated. They watch their government spend $26 billion in 1 year to support agricultural prices and ask why they can't get a little help for their essential commodity.

Another observation, though, has to do with the distortion that accompanies this frustration. Tariff support is becoming exclusionary. Nothing else can help producers, say the tariff's most passionate advocates. This could get dangerous.

The oil business took a lot of maligned hits when times were better. The "windfall profit" tax leaps to mind. Then there's loss of all but a snip of the percentage depletion allowance and erosion of the current-year deduction of intangible drilling costs.

Correction of these and other government mistakes could, as a matter of fact, help oil producers considerably. It's fair to say producers need market improvement above all else. But with WPT still on the books, 70% of a future revenue gain from old oil will go to the federal treasury. To insist, in pursuit of a tariff, that regulatory improvement doesn't count risks foreclosure of the chance for any improvement at all. Producers need both kinds of help.

The risk looms larger in the context of national politics, the last observation. Politicians outside oil country, who far outnumber their oil state colleagues, respond to T word talk with a word of their own. You might call it the N word.

No.

Houston landmarks

L.R. Aalund **April 6, 1987**

Advance parties are in Houston checking out facilities and final arrangements for the World Petroleum Congress and Offshore Technology Conference that will be held concurrently here the last week of this month.

Depressed industry conditions will certainly dampen attendance at both events, but there will still be thousands of out-of-towners in Houston. Of course, on any given week there are thousands of out-of-towners here. Many have chosen to remain permanently.

But these two petroleum industry events will bring in foreign visitors who know Houston well. If they haven't been here in a couple of years, they will find a couple of famous institutions newly shuttered.

Those places filled two basic needs: eating and sleeping.

The closure and sale of the 37-year-old Shamrock Hilton Hotel to the Texas Medical Center were reported last summer. Because this is a story that probably would not make the front page or evening news outside the Gulf Coast, it is repeated here for Oil & Gas Journal readers around the world who have fond memories of this spacious hotel and its somewhat balmy resort atmosphere.

The Shamrock housed and refreshed thousands that came to Houston for the frequent — sometimes huge — refining, production, petrochemical, and offshore meetings held there. In some cases, the grand hotel with its truly olympian swimming pool could accommodate the entire meeting.

The closure saddened many Houstonians who had grown up with the Shamrock since its builder, Glenn McCarthy, opened it in 1949 with most of Hollywood there for the biggest party Texas had ever seen. What really upset those Houstonians was the later news that this landmark would be torn down. After all, 37 years is hardly any age at all. They had assumed it would be merely converted into teaching, research, and hospital facilities.

A protest movement to stop the demolition flared briefly, but sober reflection supported the rationale of an expansion of the world famous medical center into seven or eight new high rise laboratories and medical buildings in a park-like setting. This spring the Shamrock is still standing, but its insides are gone and the outside will follow.

The other famous institution closed was the San Jacinto Inn at the Battlegrounds near the Lynchburg ferry landing about 20 miles east of downtown Houston.

This seafood restaurant had a policy of serving everyone all the boiled shrimp, crab, oysters in season, fried chicken, and biscuits they could eat for a fixed price. The food was served by fast, energetic waiters in a large, rustic dining room.

You had to assess your guests fairly well before taking them to the San Jacinto Inn. Oyster shells and crab claws were prone to fly around the table in certain crowds.

The place had been in business 70 years when it closed month before last. According to the managers, the depressed Houston oil economy played a role in the closure.

Not just an oil mud

John L. Kennedy **April 27, 1987**

It has been a long time since it was enough to define an oil-based mud as a drilling fluid "having oil as the continuous, or external, phase and water as the emulsified, or internal, phase."

Additives have long been used to adjust certain properties of oil muds. Even toxicity testing of some drilling fluid components was done at least as far back as the early 1950s.

But since the 1970s, when governments began to restrict how oil muds were used offshore, they have changed dramatically. A special cuttings handling system aimed at meeting new regulations in the Dutch North Sea was used in the early 1970s where oil muds were used to combat hole instability in salt zones.

Oil-based muds have long been almost the only solution to some difficult drilling problems — shale sloughing, salt flow, high temperature.

If it were not so, operators would have abandoned them long ago.

And certainly, if it were not so, few operators would go to the effort described in one article in the Journal's special Offshore Report. The article details Esso's use of an oil-based mud in the Beaufort Sea, an oil-based mud for which old definitions are far from adequate.

Esso summarizes the three main issues in complying with Canada Oil and Gas Lands Administration guidelines for the use of oil-based drilling fluids: choice of the base oil, oil retention on cuttings, and occupational health and safety.

Selection of the base oil involved consideration of more than a dozen properties, including its performance in the standard 96-hr LC50 toxicity test. Some properties were related to regulatory requirements. Others were important for operating reasons.

Oil retention on the cuttings is important because it helps determine whether cuttings can be disposed of into the ocean.

Special training for rig crews in operations, use of protective gear, and kick detection had to be part of the operating plan.

And before drilling with an oil-based mud, the rig's solids control system and ventilation system had to be modified and special equipment installed.

It sounds like a lot of trouble just to use a specific drilling fluid. And trouble in arctic offshore operations is spelled "m-o-n-e-y."

Why bother? Why not use another fluid and accept slower penetration or more nonrotating time?

Because, the authors say, "Oil-based mud . . . provided the only low-risk solution to a directional drilling challenge . . .

"The virtual elimination of hole problems and the more than doubling of the average penetration rate . . . provided incentive to use oil-based mud. . . ."

This report shows oil muds are still too important a drilling tool to be abandoned. And their use can meet tough environmental and operating specs.

But these days, when you say oil mud, you have to be specific. What is its aniline point? Did the LC50 toxicity test prove it to be nontoxic?

Does it contain any higher-molecular-weight polycyclic aromatics (4 to 7 rings)?

The joy of discovery

Bob Tippee **May 18, 1987**

Petroleum journalism exhilarates even its crustiest practitioners when somebody finds significant amounts of oil or gas in some new place.

The upstream side of the business has an elemental appeal: the power of human ideas, the rugged justice of risk and reward, the technical and logistical challenges of developing what's been found.

Lately, of course, Oil & Gas Journal editors have tended to greet news of discoveries with a wry, "Just what the world needs — more oil." But the sarcasm fools no one. The world does, in fact, need oil. Surpluses don't last forever. Today's discoveries will be critical tomorrow — well, someday anyway.

Surplus or no, a good discovery story makes any oil writer smile.

The surplus, however, has made many people forget the importance of future supply and the economic benefits of petroleum activity. In parts of the U.S., folks don't even want to see an oil rig.

So it's refreshing to come across a discovery story where the event seems to have pleased everyone involved.

The story comes from Bangladesh. Late last year, Bangladesh Oil & Gas Minerals Corp. (Bogmc) tapped heavy, waxy crude in a gas-producing region — the country's first oil find.

Bogmc 7 Haripur gauged 600 b/d during tests. No, that's not world-scale. It may yet prove noncommercial.

But in an impoverished country that spends 30% of what it earns in international trade to import the 32,000 b/d of crude and 12,000 b/d of products it needs, a 600 b/d find means much.

The popular effects weren't lost on OGJ's correspondent in Dhaka, who reported that the discovery inspired "euphoria."

When 7 Haripur yielded its first oil, he said, geologists on site "had an outburst of joy: 'We have it, we have it'. "

After 12 days of tests, Bangladesh Pres. Hussain Mohammad Ershad visited the drillsite, 18 miles from Sylhet.

"The people of tiny villages collected around the well," OGJ's correspondent reported. "It was a stampede, with the president and ministers all looking at what it was. Pres. Ershad was 'extremely happy' and thanked God for the deposit."

The joy of discovery. Maybe that's what the world needs more of.

Prudhoe's 10th

Bob Williams **June 15, 1987**

Alaska's oil sourdoughs tend to measure milestones in decades.

For example, in 1957, a well drilled near Swanson River on the Kenai Peninsula ushered in the modern petroleum era for Alaska and helped nudge the territory toward statehood 2 years later.

Then in 1967, ARCO and Exxon drilled the first well to tap Prudhoe Bay field. It was followed a year later by a step-out that confirmed the size of North America's biggest oil field.

On June 20, 1977, the first barrel of Prudhoe Bay oil moved into the Trans-Alaska Pipeline System (TAPS), destined for tanker loading at Valdez. That makes this week the 10th anniversary of commercial U.S. arctic oil production and the remarkably smooth operation of TAPS, history's biggest engineering feat by private industry.

Oil & Gas Journal salutes the personnel of the operating companies and contractors who have made Prudhoe Bay field and TAPS paragons of an enlightened attitude of business: operating responsibly in an environmentally sensitive area. Those operations resulted in production of 5 billion bbl of oil, half of the field's reserves, a mark reached last March.

This year marks another milestone: the last major sealift of equipment bound for current projects on the North Slope. Part of that situation stems from the oil price collapse, part from the environmentalist lobby's political clout.

The latter can be seen in yet another 1987 milestone: the start of heated debate over opening for leasing the Coastal Plain of the Arctic National Wildlife Refuge (ANWR), the best chance to replace declining North Slope production in the 1990s.

And this year offers anew the specter of a Middle East oil supply disruption as the Persian Gulf war grows hotter.

Few industry officials believe ANWR oil could be brought on stream before 2000. Remember the environmental furor over TAPS? It took an energy crisis and an act of Congress to give the country's biggest energy venture a green light.

Perhaps decades from now a descendant of mine will celebrate in print a 10th anniversary of ANWR flow. Let's hope he or she isn't writing it on a manual typewriter in a cold, dimly lit office reached by bicycle.

A year's tale

Warren R. True **July 13, 1987**

Last year was one most of us will remember as the year of the industry's worst depression.

Gas processing, along with other industry segments, has been in a slump, too. This year's Gas Processing Report shows that 1986 throughput and NGL production were off from a year earlier.

The industry in general has been hardest hit in the U.S., where the slump has wrecked several states' economies.

How bad has it been?

To help tell the tale, the American Petroleum Institute recently pulled together some 1986 statistics for major U.S. oil states.

• Among oil-producing states, Louisiana led in unemployment rates with 13.1% of its population out of work. That's 261,000 people.

Alaska had 10.8% (28,000) of its work force out of jobs in 1986. Texas had 726,000 (8.9%) out of work, second in number only to California with 892,000 (6.7%).

• Oil industry business failures — companies mainly in Texas, Louisiana, Oklahoma, and Arkansas that closed their doors with losses to creditors, according to Dun & Bradstreet — exploded by 59.9% last year to 12,704 from 7,943 in 1985. Without the drag of Texas, Louisiana, and Oklahoma, Dun & Bradstreet estimates, total U.S. business failures for 1986 would have fallen by 2.1%.

• All types of bankruptcies — Chapters 7, 11, and 13 — rose in oil states. Texas had 32,442 bankruptcy cases filed in 1986. That is a 57.4% increase from 1985. In Alaska, the increase was 66.2%, in Oklahoma 55.9%, Colorado 55.8%, Louisiana 46.6%, New Mexico 41%, naming just a few major oil states.

• The number of banks that failed in oil field states steadily increased in

1984-86, including banks that received assistance from the Federal Deposit Insurance Corp. (FDIC) that otherwise would have failed. Twenty-six failed in Texas last year, 16 in Oklahoma, followed by Kansas with 14. For the U.S. as a whole, 145 banks failed last year, including seven FDIC assistance transactions.

Banks are failing this year at a more rapid pace — more than 30 in Texas alone.

Yet some signs point to 1987 as a turnaround year.

Slowly rising oil prices suggest a recovery may have begun. And start-up last year of two large gas processing plants has helped return some optimism to gas processors.

Conoco Inc. brought up its San Juan (N.M.) plant with a production capacity of 45,000 gal/day. And ARCO Alaska brought on stream its 3.3 bcfd Prudhoe Bay plant. As a result, U.S. and non-Communist gas processing capacities rose for 1986.

Another, less tangible sign of a turnaround: On Houston's Southwest Freeway a convoy of three flatbed trucks loaded with drill pipe was recently spotted heading out of town. A rare sight in recent years — and a welcome one.

The organic way?

Bob Williams **July 20, 1987**

One holdover of the 1960s mentality is the yuppie/boomer's (YB) unquenchable penchant for things "organic."

Like anything the focus of cultish mentality, use of the term organic soon grew far beyond its dictionary definition. Essentially, organic came to be interchangeable with "natural," i.e., unsullied by human touch or manufacture (UHTM).

Another holdover of the '60s is the YB's perpetual yearning to quest through the wilderness, especially through a wilderness UHTM. In fact, wilderness came to mean a place UHTM. Other definitions need not apply, thank you. (Attention, petroleum industry: This means especially you).

Somehow the notion arose that petrochemicals weren't "natural" in origin. Any tome on organic gardening shuddered at the thought of petroleum derivatives sprayed on your Whole Earth Peace and Light Garden. Never mind that there is a scientific distinction between inorganic and organic chemicals — and guess which falls into the latter category.

But there's a blurring of meaning lately, for anyone, like the undersigned, long a daisy-sniffing backpacker. When hiking first took off in the 1960s, gear tended to be metal and leather and wood and down and cotton, etc. — and heavy and bulky and uncomfortable.

Enter an outdoors outfitter these days, and everywhere you'll see testimonial to the virtues of petrochemicals: light, waterproof, breathable, durable, flexible, comfortable. An outfitters' catalog and much of the trail talk nowadays is devoted to singing the praises of polypropylene, polyethylene, polycarbonate, urethane, sorbathane, Velcro, Gore-Tex, Neoprene, Nalgene, EVA form, Vibram, Thinsulate, Quallofill, Cordura, Lexan, among many petrochemical products. Pretty pricey stuff, too.

And the YBs can't get enough of the petrochemical-based wilderness gear, much less all the wondrous new gadgets for their all-terrain bikes, ATVs, VCRs, CDs, Volvos, sporting togs/accessories, and the like.

Consumer demand for petrochemicals in the U.S. is strong and is expected to continue growing, thanks in part to lower oil prices, as can be seen in our report beginning on p. 17. Continued growth will hinge to a great extent on whether oil and gas prices stabilize in a range of $20-30/bbl of oil equivalent.

How well the U.S. reins its dependence on oil imports could be a critical factor in whether there will be more oil price spikes in the future. And industry officials agree that federal land access is pivotal to reducing that dependence. If the oil price talk in the 1990s presages $40-50/bbl, petrochemical products could get even more pricey.

Our YB friend might bear that in mind for the next shopping trip to the outfitter's to gear up for a UHTM wilderness trek. Chances are he'll pay for it with one of those trendy new credit cards that earmark part of the fee to righteous causes such as Greenpeace and the Sierra Club.

Like any YB, he'll pay for it with plastic.

A job with a future

L.R. Aalund **September 28, 1987**

If your son or daughter or a friend is entering a university as a freshman this fall, would you, if asked, advise him or her to study petroleum engineering?

That question could stump many industry professionals, given the major, lingering slump in the world oil industry and the devastating effect this has had on careers, personal finances, and some regional economies.

The answer would be easier if you were addressing a Saudi student rather than one from Texas, Oklahoma, or a half dozen or so other states and Canadian provinces. But the department of petroleum engineering of the

University of Texas at Austin is unshaken in its enthusiasm for petroleum engineering as a course of study.

In a long letter last January intended for high school seniors and counselors, Prof. Gary A. Pope, chairman of the department of petroleum engineering, wrote in essence that it is an excellent year to begin studying petroleum engineering.

He used historical UT job-opportunity charts, which he said offer clues to the future, to make his point. They show that in general as the number of wells drilled increases, the number of job offers increases.

But even during 1969-72, the bottom of the previous petroleum industry down cycle, UT petroleum engineering grads received an average of about two job offers each and essentially 100% found jobs. This was because only a handful of students elected earlier to study petroleum engineering. As the chart also shows, 1977-81 was a very good time to graduate but a bad time to start work toward a degree. In 1983-84 there were two students for every job offer.

The less than 20 petroleum engineer graduates Pope expects at UT in 2-3 years will have more than enough job offers, based on the fact that there were 78 job offers in 1985-86 when the industry was heading for the bottom. As other incentives, he lists oil company support, scholarships, high-paying summer jobs, and the increasing need for technically advanced personnel.

This is all pretty convincing to a sharp student — unless the tuition payer in his or her family is a laid-off, 50 year old, early retired petroleum engineer beginning a new career in aerospace or real estate.

Independent focus

Jim West, Alan Petzet **October 12, 1987**

As you can see in this week's special report, U.S. independent operators generally are recovering from the bloodbath of 1986.

Most of those lucky or astute enough to still be around are recovering, that is.

Lots of them have faded into Chapter 11 bankruptcy, possibly never to be heard from again. Others definitely will never be heard from again. Fine old independent companies like Aminoil USA Inc., Natomas Co., Pioneer Corp., and General American Oil Co. of Texas were acquired in recent years by other companies as the industry restructured in response to changing times.

For most independents, the changes of 1987 are a welcome relief from the disaster that was 1986.

Still, a few keep their sights on severe problems of the recent past. That's what Independent Petroleum Association of America Pres. H. B. (Bud) Scoggins Jr. told producers in Oklahoma City last month.

"No place is the attitude of the people in the industry so negative as it is here in Oklahoma, particularly here in Oklahoma City," Scoggins said in a pep talk at Oklahoma Independent Petroleum Association's annual meeting.

"No place else are people so focused on the past and what happened last year and spending so little time thinking about the here and now and looking forward to the opportunities of the future. That's really sad."

Reports of better times come from many producers.

In the Appalachian basin, for example, Petroleum Development Corp., Bridgeport, W. Va., was able to fund and drill only four wells during the first 9 months of last year.

But it drilled nine wells in the first 6 months of this year. And last August it closed a program calling for participation in about eight wells to start immediately.

"We have not been able to close a midyear program of this sort for 4 or 5 years," said PDC Chairman James N. Ryan.

"We believe this success reflects the strength of our programs and a turnaround in investor interest in oil and gas drilling."

However, there's no denying that there still are many uncertainties about the times.

As Scoggins told OIPA, "All of us have been subject to tremendous pressure because of what's happened, but keep in mind that a diamond is just a piece of coal that made good under pressure.

"Let's turn that around.

"Let's put the focus not on our problems but on the nation and the consumer.

"Only if and when we can convince others that it's their problem we are trying to solve are we going to get any help out of Congress, the federal government, or the public."

Motives

Bob Williams **October 19, 1987**

Oil & Gas Journal gets a wide array of letters.

Most are from industry types that are either corrective on technical points, laudatory on content, or argumentative on an opinion.

Then there are the ones that are written on Big Chief tablets with Crayolas from someone whose tooth fillings are picking up radio signals from Antares telling them that the Trilateral Commission is really behind Big Oil's conspiracy to fabricate another energy crisis and are-you-guys-gonna-spill-the-beans-or-are-you-part-of-the-conspiracy?

Now and then OGJ gets letters from perfectly rational, intelligent readers in the industry that just result in puzzled head-scratching. That's the case with letters, requested not to be published, charging accuracy and bias in the same sentence in coverage of Offshore California: to wit, why doesn't OGJ write upbeat news about Offshore California, not just about the controversies? That just scares away companies who otherwise might spend money off California.

Thanks. If OGJ is doing its job, it reflects what's happening and what's not happening and why. Right now, there is more that is not happening — in light of its potential — off California than is happening. For every Platform Irene that gets installed with relatively little difficulty, there are three horror stories: Gail, Santa Ynez, Coal Oil Point.

The letter writers admit their motive: losing prospective work because of what a company reads in OGJ. But accurate coverage helps our readers make the right — even if negative — business decision.

Access to acreage off California is usually defended by industry as necessary for U.S. energy security, but also as good for the economy. That's when the environmentalist lobby and its political hit men jeer that the industry is motivated by "obscene profits."

Any businessman will admit to a profit motive. That's why he's in business. But impugning a profit motive by environmentalists in a previous column netted me this criticism: You shouldn't second-guess someone's motives.

Enough said. Could one presume other than that the horde of lawyers that descended on tiny Alaskan villages to sue in their names to block Bering Sea sales only wanted to preserve a zillion acres of whale pasture they'd never seen? Or that former Alaskan officials who blocked leasing in Bristol Bay were overlooking profits from the interests they held in fisheries there?

Is it fair to think that the big bucks industry pumps into the fisheries enhancement fund off California is for anything other than enhancing the welfare of fish? Or that Sierra Club, Greenpeace, Wilderness Society, et al. are guarding a sacred trust by railing against offshore drilling and ANWR leasing and not just maintaining a high profile to keep the subscription rolls and donation coffers brimming?

Is it right to suspect that the coauthor of California's Proposition 65, governing discharge of suspected carcinogens and teratogens in or near water supplies, is anything other than public-spirited by later offering his services as a private attorney to companies to represent them in complying with Proposition 65?

Naaah . . .

Gasohol questions

Bob Tippee **January 4, 1988**

Raising questions about gasohol can be as perilous as a jog in the smog.

When you express doubts about blended-fuel programs you must prepare for charges that you are a biased curmudgeon with your facts crooked. If only you believed what people trying to sell gasohol said about gasohol you'd see that motorists want the fuel but are being tricked into not buying it by biased curmudgeons with their facts crooked. Then you would not question government efforts to force motorists — who, you must remember, deep down want to use gasohol — to use gasohol.

You also would ignore your silly suspicions. You would not, for example, wonder whether gasohol promoters were hitching their wagon to environmentalism because their product, even with state and federal subsidies, faltered in the market.

What the heck? Some questions need to be asked.

And this question focuses on Denver, a city with an air quality problem. Beginning this winter, gasoline sold in it and other Rocky Mountain Front Range cities must have oxygen concentrations greater than normal. This, it is believed, will reduce auto emissions of carbon monoxide, one of the region's troublesome pollutants.

A coincidence of regulations and progressive tightening of the oxygenate standard soon may make gasohol the only salable fuel in the affected cities. This works out splendidly for gasohol producers, who support the endeavor all the way — for environmental reasons, of course. For motorists who don't wish to use gasohol — well, tough. Clean air, you know.

But wait. What if there were a way to clean the air without regulating away consumers' motor fuel options? What if people drove less?

Toward that end, Total Petroleum Inc. is supporting a Denver campaign asking commuters to leave their vehicles at home 1 day/week. Motorists filling their fuel tanks at Total's Vickers stations receive coupons for free round trips on Regional Transportation District buses.

Sure it's an effort to sell more gasoline. That's what Total's in business to do. But the intent, to restrict use of the company's product for the sake of air quality, gives the program a twist that makes it something more than a sales gimmick.

Who knows? It might help make Denver's voluntary no-drive program a success. And that might clean the air to the point that involuntary use of gasohol no longer made sense, if it ever did.

Total has said nothing to link its no-drive support to blended fuel mandates in any way. But what if driving limits did reduce Denver's carbon monoxide pollution to tolerable levels? Would gasohol proponents withdraw their support for a program narrowing consumer options to a fuel more expensive than the proscribed — and, some say, higher quality — alternatives?

Total isn't posing that question.

The biased curmudgeon with his facts crooked is.

A pioneer contract

John L. Kennedy **February 1, 1988**

Propane futures trading has had a hard life. But its time may now have come.

One month short of 20 years after first being inaugurated, it got another chance last August on the New York Mercantile Exchange.

Selling the concept was tough in the mid-1960s. Harry Thomas, a Tulsa consultant, was instrumental in establishing the original propane futures contract. He conducted seminars and met with individuals to explain how futures trading might help propane buyers and sellers.

One key element that makes a futures market work was certainly there — price volatility. But other key elements were missing.

Speculators, for example. The industry wasn't quite ready to accept speculation as a legitimate part of the market.

It wasn't that buyers and sellers didn't speculate; they couldn't avoid it. They just didn't call it that. And whatever it was that they were doing, it turned out that they wanted to do it in the cash market instead of a futures market.

Nor was hedging that appealing. It meant you limited your chance for profit as well as your risk of loss. There was just something about speculation, hedging, and "futures" that didn't seem right.

Things are different now. The concept, at least, is accepted. Roller coaster crude prices have made believers out of many crude oil and products traders.

As a result, trading in energy futures contracts has shown impressive growth, as our special report in this issue details. In 1987, 24 million energy contracts were traded at Nymex, for example, compared to 13 million the year before.

When OGJ's May 1, 1967 article explaining how a propane futures market would work was written, no one gave much thought to crude price fluctuation.

Propane was different, of course. A Journal headline in late 1967 reported, "Propane overcorrection to bring big spring surplus." During earlier winters, supplies were dangerously low.

The propane market seemed a natural for hedging.

Indeed, when Commodity Exchange Inc. inaugurated propane futures trading in September 1967, trading volumes in the first weeks ran at more than 60 contracts/day. But by the end of the year, volume had settled back to 5-20/day.

Then curbs on trading were necessary in 1968 because of delivery problems. Actual deliveries were much higher than expected, reflecting industry's reluctance to deal in anything but the physical market.

In early 1972, LP-Gas Associates of the New York Cotton Exchange adopted an improved propane futures contract.

But apparently the time still wasn't right.

Now it may be.

The propane futures contract that began trading in August is apparently off to a good start. It promises to become a useful tool, and the proposed natural gas futures contract should complement it.

But it hasn't been easy.

P. C. Lauinger

Gene T. Kinney **February 29, 1988**

The year was 1931. Times were hard for Oil & Gas Journal and the industry. Taking over a family-owned petroleum publication wasn't exactly the most promising prospect for a young man.

Frank T. Lauinger, president of Petroleum Publishing Co., had died unexpectedly, and the legacy of Journal leadership fell to his son, P. C. Lauinger. P. C., as he was to become known throughout the industry, had served as aide to his father for 9 years.

He had prepared himself for the top job. But thrust upon him so suddenly, in the midst of a depression, it was an awesome prospect. A publishing giant was eager to take the magazine off his hands.

The temptation to sell was strong. But he had faith in the industry and the Journal and decided to try to make a go of it himself.

P. C. Lauinger died last week, knowing that his faith had been vindicated many times over the 57 years since making that decision.

When P. C. decided to take the job, he moved to Tulsa, planted his feet firmly in the oil patch, and set about building a technical staff to serve his petroleum industry audience as it moved from the rule-of-thumb phase of its history into the new era of technology.

And he did more than just make a go of it. His secret, if he had one, was that he always knew who was boss. It was the reader.

P. C. inherited his dedication and fierce integrity from his grandfather,

Patrick C. Boyle, an ex-scout for John D. Rockefeller who had built the Oil City Derrick into the first true petroleum publication. When Boyle founded Petroleum Publishing Co. in 1910, the industry was sharply divided by chaotic conditions and the bitter battles with Rockefeller's Standard Oil Trust. He set out to instill an industry consciousness into the warring factions, preaching unity and the need to think of the good of the industry as a whole.

An active oilman as well as a publisher, P. C. took that concept and refined it to a higher level of service, providing information vital to oilmen in the conduct of their business.

P. C. was always an editor's publisher. Convinced that editorial excellence was the route to business success, he was single-minded in his pursuit of that goal. Guided by that philosophy, he made Oil & Gas Journal into the "bible of the oil industry."

But his greatest success was not as a businessman or philanthropist, community, civic, or religious leader, although he certainly was all of those. It was his humanity that set him apart, his relations with individuals. Employees who traveled the world in their jobs grew accustomed to finding friends of P. C. wherever they went. They always asked about him. It was a personal question.

To staff members, business associates, and others with whom he came in contact, he was leader, father figure, mentor, friend, loyal supporter — a revolving gentleman who showed the same gentility and warmth at every turn, every hour of the day.

In recent years, under the leadership of a son, Philip C. Lauinger Jr., the company expanded with P. C.'s blessing into other publishing fields. But his heart was always in the oil patch, with Oil & Gas Journal and the oilmen readers he sought to serve.

No more booms, please

Bob Tippee **April 11, 1988**

As forms of expression, bumper stickers rank somewhere below puns and greeting cards. However clever or insightful the message, the medium dooms it to overexposure.

Consider the current chortlings of U.S. Northeasterners who did not, as a handful of bumper stickers suggested a few years ago, freeze in the dark. The sentiment — itself a response to OCS leasing opposition in the Northeast — got more exposure than it deserved. And the reason for the exposure has more to do with the message's incivility than with the number of vehicles it adorned.

Less offensive, thanks to an implicit humility, is a successor that pleads to the Almighty for another oil boom and promises, this time, not to squander things one way or another.

Oil industry speech writers love it — the delicate versions, anyway. Who, by now, hasn't had to dredge up a chuckle for one more banquet toastmaster who acts like he saw the sticker for the very first time on a car parked outside the luncheon hotel?

"Please, God, send me another oil boom. I promise not to —" etc. Har-har.

Yet there may be something to this. Overexposure might mean something in this case.

Do oil people who have survived the mid-1980s really want another boom? Surely not. To hear the survivors talk, the best hope is for turnaround followed by stable, steady growth in petroleum demand, prices, and industry activity coupled with certainty and fair play from the government.

So what's the attraction of the bumper sticker prayer? Why do speakers keep quoting a joke everybody's heard at least five times?

It must be the allure of the unattainable. Nobody expects another boom. Nobody thinks it would be good for the industry. But wouldn't it be nice if . . .?

Well, would it?

Having experienced a depression no one in the 1970s could have imagined, what would industry do with another flush period like 1980-81?

Diversify risk? Protect liquidity? Moderate debt? Hedge in the futures market? Resist the assumption that the future will be just like the present only more so? Move to Minnesota? Marry a banker?

It's worth wondering about. Figuring out how to handle a boom that probably won't happen might generate ideas helpful in the modest recovery that more likely will.

And here's a suggestion: No matter what comes next in the oil business, let's do whatever gloating may be necessary in private, not on our car or pickup truck bumpers. Things change.

Say, that's catchy: Things change. It might look good on a T-shirt or baseball cap.

Period

L.R. Aalund **April 18, 1988**

It was perhaps fitting that Oil & Gas Journal's technology editor covered the final event of the Pennzoil/Texaco epic. His primary responsibility is not business news but to see, with some help, that 12-32 pages of this magazine's technology section are filled with quality articles each week.

The final act of Pennzoil vs. Texaco, a press conference by Pennzoil Chairman J. Hugh Liedtke after Texaco had transferred $3 billion to Pennzoil's bank account, was rather anticlimactic. No one believed Texaco would fail to fork over the money, the cash did not come in armored cars, and the former foes conducted themselves as if they had just settled a slight business disagreement.

So we really didn't need anyone there who had worked on this story and its every nuance for more than 4 years, as our Texaco/Pennzoil specialist, Rick Hagar, had. That was lucky because he was at a business lunch at an unknown location when we were informed of the surprise press conference 45 min before its start at 1 p.m. in downtown Houston, about 8 miles from OGJ's office.

Photo opportunities there were poor. One showing a group of managers huddled around a monitor in a Houston bank watching the $3 billion blips come in and another of Liedtke holding a letter from the bank saying the funds had been received will hardly be Pulitzer candidates.

Our first impulse was to just forget about it, knowing that it was all over but the shouting. But the thought occurred that Pennzoil could announce something big, like the purchase of Texaco's Port Arthur, Tex., refinery with part of the money. Also, OGJ, after spending so much ink and paper on this story, would have to put a period on it.

The press conference on the 33rd floor of Pennzoil Place was in a rather elegant conference room cluttered with reporters, spectators, cameras, wires, and photographer's satchels.

The conference started on a dramatic note when one large photographer slipped while snapping pictures and trying to stay ahead of the Liedtke entourage as it marched into the room and fell with a thud that seemed to shake the Pennzoil Place skyscraper. He quickly bobbed up with a rather forced smile, momentarily not snapping but apparently unhurt.

There were then some good questions by the assembled print and electronic media and then some, well, some People Magazine-type questions:

Q. Mr. Liedtke, where were you and what did you say when the fund transfer was made this morning and you knew you had the money?

A. I was correcting a letter to a friend of mine concerning painting a fence.

Q. Are you speaking with Messrs. Kinnear and DeCrane (Texaco top executive and chairman, respectively).

A. We speak. Obviously we are not at all close.

Q. How much would you help George Bush become president?

A. In any way I could.

Q. Would you consider becoming secretary of Energy if he is elected?

A. . . . no.

EOR: 1988

G. Alan Petzet **April 25, 1988**

About every other year since the early 1970s, OGJ editors have covered a major conference on enhanced oil recovery in Tulsa, Okla.

The magazine even co-sponsored the meeting a time or two. Sponsors these days are the Society of Petroleum Engineers Mid-Continent Section and the U.S. Department of Energy.

The vast potential of EOR and the dedication of EOR people have not changed since the early meetings, but the economics of EOR, as dictated by oil prices, have been on a rollercoaster ride for nearly 15 years.

In 1971, according to OGJ's EOR survey, the U.S. had 132 EOR projects. Most were small pilots, and producers received an average of $3.39/bbl for their crude that year.

Oilmen came to the Tulsa meeting to deliver the message of EOR's promise: The U.S. has produced 150 billion bbl of the 450 billion bbl of oil in place, leaving 300 billion bbl as a target for EOR. Recovery of 10% would double U.S. oil reserves.

Later in the 1970s, oil price spikes came along, and the number of U.S. EOR projects grew.

History's highest oil prices in 1979, decontrol in the U.S. in 1981 and a federal incentive program combined to hasten EOR's progress from laboratory to field.

The program allowed producers to prematurely charge market prices for price-controlled crude to pay front end costs. By 1986, OGJ's count of field projects swelled to 512.

The 1988 survey, published Apr. 18, lists 366 projects, down 28% from 2 years earlier.

DOE is studying whether an exploration tax credit could be applied to EOR projects, but Energy and Interior Department officials at the 1988 symposium last week indicated that the industry shouldn't count on further federal assistance for EOR.

Another speaker said the federal government and other states should adopt tax incentives, perhaps patterned after an Oklahoma law that takes effect July 1.

That law generally exempts incremental EOR production from qualified projects from the state's 7% gross production tax until payout or 36 months, whichever is first.

The state doesn't contribute existing revenue at first because only the oil it would not have had otherwise is exempt. After the exemption ends, the state's tax stream increases.

One question: Which is the legacy of the 1979 DOE front end incentive program? Is it the period through 1984 when the number of EOR field tests grew, or is it the years since then when field work waned because no artificial incentive was available?

An OGJ editor and his wife expect an addition to their family next month. The obstetrician and hospital want all the front end costs paid in advance, years before the child is likely to produce any financial or other return.

For the time being, it looks like the challenge for EOR people may be to raise EOR to maturity using their technical expertise without help from much higher oil prices, inflation, or incentives.

Gulf of Mexico State No. 1

L.R. Aalund **May 2, 1988**

Fifty years ago this spring, a once well-known company completed the first oil well in the Gulf of Mexico.

This anniversary, which probably took place in April (drilling was completed in mid-March), hasn't been marked by any fanfare. For one reason, the two companies behind the project have disappeared into Mobil and Unocal. This anniversary would in truth have escaped our notice had not this editor come across one of the first articles he had researched a good many years ago. The 25th anniversary didn't, to our knowledge, get much notice in 1963 either.

Historical ramifications aside, it is still a neat story that shouldn't be allowed to fade away. We owe many of the details to a company house organ, the Pure Oil News, which reported in its May 1938 issue:

"Considerable interest is being manifested in the recent oil discovery 1 mile out from the Louisiana shore."

The well was drilled in 14 ft of water from a platform mounted on piling about half way between the mouth of the Mermentau River and the then tiny port of Cameron at the mouth of the Calcasieu. Cameron is now a major offshore supply base and fishing port. And the mouth of the Mermentau is now, courtesy of the U.S. Army Corps of Engineers, some 8 miles east of where it was then.

The writer goes on to report, "The discovery well is known as the Gulf of Mexico State No. 1, owned jointly by Pure Oil Co. and Superior Oil Co., the latter company being the operator. The well . . . is located on a 40,000 acre block of leases known as the Creole Field."

Then the author prophetically wrote, "This discovery on a large block naturally is heralded as a pioneering project of the highest order."

That part of Louisiana, with its vast sweep of marshland, was practically inaccessible in 1938. The fur trapper's "pirogue" had been the primary means of transportation — until the marsh buggy appeared on the scene and took oil explorationists into the home of the muskrat and otter. Only recently had the first road between Lake Charles and Cameron been built and the steamboat service on the Calcasieu River discontinued.

Although only a mile out, the rig was more isolated than any in the gulf today. There was no radio and only a one-line telephone system, which was usually out of commission, between Cameron and Hackberry. All the heavy equipment went out on flat-topped barges towed in some cases by chartered shrimp boats, which also carried the drilling crews. Old pictures show service company trucks on those barges. During fog, the boat skipper, if he thought he had gone far enough, would shut down the engine and everyone would go to the railing to try to hear the rig. There were sometimes differences of opinion as to which direction the sound was coming from.

The well was drilled to some 9,300 ft, plugged back to 5,610 ft, and perforated. The platform was then enlarged, and 10 more producers were drilled directionally.

On second thought, perhaps this golden anniversary was celebrated in a fitting fashion last month when Amoco announced it had drilled an oil and gas discovery in 3,560 ft of water in the Gulf of Mexico.

Willing buyer, willing seller

Jim West **June 20, 1988**

It's easy for those of us who live in oil country to harbor the same provincialism as those who live elsewhere.

We can become so engrossed with reporting industry activity we sometimes forget the role of a key player in the supply/demand equation: the consumer.

It isn't exactly a secret that all of the world's oil and gas — found, produced, and processed at no small expense — has to go somewhere to be consumed. So there's bound to be nothing as frustrating as a roadblock between consumer and producer. The results are a willing buyer unable to buy and a willing seller unable to sell.

Matthew C. Cordaro knows the willing buyer's frustration. He spelled it out for the Federal Energy Regulatory Commission during a hearing on proposed gas pipelines to serve the U.S. Northeast.

Cordaro is president of Long Lake Cogeneration Corp., a subsidiary of

Long Lake Energy Corp., New York. Among his company's projects are five Long Island cogeneration plants with a combined capacity of 500,000 kw. All are designed to rely on gas as their main fuel but will be able to burn No. 2 fuel oil as a backup. Another Long Lake project, also on Long Island, is a 238,000 kw oil-fired combustion turbine that ultimately could burn gas.

With Long Island power demand showing robust growth, Long Lake's plans sound like a formula for prosperity. As its units go on line, their total gas demand could rise to 133 MMcfd by 1993, making Long Lake the biggest new buyer of gas in the Northeast. But there is a problem. There is no assurance the gas will be available.

"We have reached a critical point in development where secure natural gas supplies must be obtained for our projects," Cordaro told FERC. Hence his "immediate, overwhelming" interest in what happens to the batch of proposals lodged with FERC to step up gas deliveries to the Northeast.

Long Island certainly could use more energy.

The local power company, importing all the electricity it can, is curtailing customers. It hasn't been able to obtain a license for its 810,000 kw Shoreham nuclear plant. Gas demand figures have been understated because, until recently, they were based on residential heating needs. They did not take into account gas to fuel added power generation.

Cordaro said, "We urge FERC to drastically accelerate its schedule for approving and implementing as many new gas pipeline projects as are necessary to serve Long Island."

Columbia Gas System Inc. shared that sentiment until last January. That's when it withdrew from a group that proposed the Erie Pipeline System to serve the Northeast. Erie in February 1987 was the first major interstate proposal filed under something FERC called its optional expedited certificate procedures.

Columbia was frustrated by FERC's lack of action on the 11-month-old application. "The fast pace of today's gas business makes it imperative that Columbia focus its resources on other opportunities that can be realized sooner," said Chairman John H. Croom.

Eighty years in Houston

Rick Hagar **June 27, 1988**

Those looking for Oil & Gas Journal's Houston office won't find it today at 1200 Post Oak Blvd., across the street from Willie G's Seafood Restaurant.

We're still in the Galleria area west of downtown, but we moved during the weekend into the Natural Gas Pipeline Building just south of Transco Tower.

Our telephone number and post office box haven't changed, but our new street address is 3050 Post Oak Blvd., Suite 200. The delivery zip is 77056.

The move represents a significant upgrading of our Houston facilities. And it demonstrates the commitment that our parent, PennWell Publishing Co., has to the oil and gas industry in general and to Houston in particular.

Of course, neither commitment is new. OGJ has been serving the petroleum industry for 86 years. We've had a Houston office for 80 of those years.

Our Houston connection began in 1908 when Oil Investors Journal opened a branch office in the Bayou City.

Holland S. Reavis, a St. Louis newspaperman, had launched the magazine 6 years earlier about 80 miles east of Houston in Beaumont, where Reavis had been drawn by Spindletop to seek his fortune in the Gulf Coast oil boom.

His magazine prospered, and in 1907 Oil Investors Journal opened an office in Tulsa, where headquarters were eventually moved.

The Houston branch office opened 1 year later in Room 229 of the Commercial National Bank Building at the corner of Main Street and Franklin Avenue.

Growth quickly caused the editors of Oil Investors Journal to seek a bigger home in Houston. Early in 1910, the Houston office moved to Room 7 of the Southwest Building at 910 1/2 Texas Ave.

That was followed a month later by a name change of the magazine to reflect its expanded coverage of the petroleum industry upon acquisition by the present owners. Oil Investors Journal became Oil & Gas Journal effective June 16, 1910, with Vol. 9, No. 1.

Oil & Gas Journal began its modern era, marked by the end of World War II, with the Houston office on the eighth floor of the Sterling Building at 600 Fannin.

The office moved 20 years later to the fifth floor of the Main Building at 1212 Main Street. Then in mid-1972 it left downtown Houston for the Post Oak Boulevard address.

OGJ today is getting itself ready for the petroleum industry's next era, which is one reason we have upgraded our Houston home and signed a long term lease at the Natural Gas Pipeline Building with several options to renew and expand.

When you're in the Galleria area, stop by for a visit and a tour of our new office.

Help us celebrate OGJ's 80 years of covering the oil and gas industry from Houston.

Call of the neodruids

Bob Williams **July 4, 1988**

Did you know that environmentalists cause cancer?

O.K., stop right there. Once and for all, let's drop the term "environmentalist." It's meaningless. Name one antienvironmentalist or, for that matter, an ex-preservationist or a nonecologist.

A better choice of terms, with acknowledgment to the great writer John McPhee, would be neodruid. The ancient druids sacrificed human beings in the worship of trees. Modern druids aren't quite so limited or so literal, although an argument could be made against that latter point for some groups.

Anyway, neodruids cause cancer. Here's the case: More than 99.99% of the carcinogens we ingest are from natural or traditional sources such as tobacco, alcohol, and cooked food. Oranges contain benzene, potatoes have arsenic, and peanut butter has aflatoxin. That's from an article in Chevron World magazine by Elizabeth Whelan, director of the American Council on Science and Health.

Well, then. Environ-, uh, neodruids are always promoting "natural" substitutes, especially for food. Ergo, they promote cancer.

Pretty stupid logic, isn't it?

Now compare that logic with this observation by the American Wind Energy Association (AWEA): The burning of fossil fuels may have triggered the devastating drought gripping much of the U.S. in recent weeks, among a host of other environmental problems.

"Despite the fact that conventional energy sources are increasingly linked with major long term environmental disasters such as this summer's drought, the Reagan administration and Congress are working hand in hand to block any major progress on promising new energy technologies like wind," says AWEA Executive Director Tom Gray.

Before he's finished, Gray trots out other environmental problems associated with fossil fuels: acid rain, arctic wilderness oil exploration, damage from offshore drilling, and the like.

Gray's climatic observations stem from testimony before a Senate environmental committee by NASA Climatologist James Hansen that the "greenhouse effect," the long term global warming attributed to atmospheric pollution, has begun.

Gray omits Hansen's disclaimer that it's not possible to attribute a specific heat wave to the greenhouse effect and that other climatologists insist warmer temperatures in recent years may be within the normal range of climatic changes.

AWEA, by the way, decries a congressional budget proposal calling for only $10 million federal program to develop an advanced wind electric turbine, which was turned down by the House and Senate.

But will farmers with scorched crops and ranchers with dying cattle — not to mention the sweltering millions — voting in November remember the alleged link with fossil fuels?

Maybe they won't recall government reports last week that sunspot activity is increasing rapidly and that more climatologists are linking solar disturbances with changes in rainfall patterns in the U.S.

And that without benefit of an environmental impact report by the sun. Let's see the solar dru-, uh, lobby defend that one.

Gas then and now

John L. Kennedy **August 15, 1988**

Despite their steady march northward into deeper water and more hostile weather over the past two decades, North Sea operators, particularly in the U.K., are not forgetting the roots of what has been one of the biggest, most expensive offshore plays ever.

They're back developing gas in the southern basin, where British Petroleum Co. brought in the North Sea's first commercial strike in U.K. Block 48/6 in late 1965.

This well and subsequent strikes, said Oil & Gas Journal in the July 4, 1966, issue, ". . . poked a hole in a hornet's nest. Everybody wants to avoid the wasps."

"Everybody" included the British Gas Council, worried that the public would pay too much for the gas; oil companies worried about losing heating oil market to gas; coal interests were concerned that a low-priced product would be a competitive threat; and smaller oil companies without fuel oil business worried that they'd have to wait too long to put gas production on stream.

Slightly more than a year later, Oil & Gas Journal's first special North Sea Report said the gas play had turned up 16 producers and 20 dry holes, and the first North Sea pipeline had been laid.

Much more drilling and development work was done, of course, in the southern gas basin. But after oil was discovered in the North Sea and the challenge of increasingly severe operating conditions brought new romance, the gas play lost some of its glamour.

Even during the glory years of oil, though, more gas was found. In 1972, for instance, when 15 oil fields were reported discovered, there also were eight gas discoveries made, including three in U.K. waters.

Our latest special report on Offshore Northern Europe in this issue is ample evidence that gas interest and gas potential are both still high — from the North Sea's southern gas basin to the Barents Sea.

In the southern basin, nine new gas projects have been approved by the U.K. Department of Energy this year alone. At least 16 other gas and gas condensate projects are on the horizon.

In the Barents Sea, preliminary drilling that indicates the area may hold more gas than oil has dampened the spirits of some hoping for big oil finds.

Still throughout northern Europe waters, gas gets respect.

Rather than fearing what new gas would do to fuel oil and coal markets and worrying about whether the consumer can afford it or not, governments and operators now are concerned with replacing supplies that will be depleted in the early 1990s.

These supplies are needed for markets that have developed in the short time since gas was first discovered in British waters.

A little over 22 years ago, there was no oil or gas production in the North Sea.

Now, one of the hot spots is where it all began.

Foraging at Cold Lake

L.R. Aalund **August 22, 1988**

Through interviews, pictures, editing, and research, Journal editors can get intimately involved with projects or complexes they have never seen.

Naturally an editor will over time build a mental image (at least this one does) of the setting of these facilities. This can lead to a few surprises when he finally gets on location — like seeing rain flooded streets and soaked camels in Abu Dhabi City. This is the desert?

Earlier this month this editor visited Cold Lake, Alberta, Canada, for the first time, an area with advanced oil recovery projects that he has followed for a number of years. It too held some delightful surprises.

The 3 1/2 hr bus trip from Edmonton to Cold Lake was half the fun. It was sponsored by the Alberta Oil Sands Technology and Research Authority as a preamble to the Unitar/UNDP heavy oil conference. Some 30 assorted North and South Americans, Chinese, Germans, and Hungarians chose to make the trip on a bus jockeyed by a driver named Doris, fittingly from the Grey Goose Bus Line.

After the steaming Gulf Coast, the coolness of northern Alberta was bracing. And there were indeed north woods, and ponds and lakes with smart ducks that will soon be winging south to Louisiana. But there were also little towns, some with exotic-looking Ukranian churches, and neat farms with big fields of wheat and oats straining to ripen before the first frosts in September.

The people who run and work the oil operations at Cold Lake, Wolf Lake, and Fort Kent therefore don't live in remote camps but in small towns like Bonnyville, Alta.

Out on the projects there are modern office buildings with curbed parking lots, training centers, and landscaped lawns. But back in the woods where clusters, or pads, of pump jacks work in solitude it was like it was supposed to be, with beaver lodges, birch, spruce, and a touch of cool stillness that tells the tenderfooted southerner to get out before late fall.

These woods took on new dimensions thanks to this traveler's seatmate on the bus, Dick McGregor of the Department of Industry in Ottawa. Dick has taught north woods survival to Boy Scouts. He is expert and articulate on the flora and fauna of those regions.

On the edge of one clearing, thanks to him, we sampled rose hips, the fruit of the wild rose that is loaded with vitamin C, red bunch berries, tender weed stems rich in protein, and clover leaf and blossoms. Had we a pocket knife, Dick said we could have tried a particularly nutritious bark.

In a way it was, as originally pictured, a bit of wilderness.

I liked Miss Efficiency

Rick Hagar **October 24, 1988**

A highlight each year of the Texas Railroad Commission's annual state of the industry hearing is the presentation from Clayton W. Williams Jr., the plain-speaking oilman from Midland.

The hearing features about two dozen speeches from top executives of various exploration/production, refining, and gas transmission companies doing business in Texas. The speeches usually are monotonous and repetitive.

Williams, who heads a company that bears his name, grew bored during the 1985 presentation in Houston. A bored Clayton Williams is a volatile Clayton Williams, as subsequent events demonstrated.

Shortly before he was due to make his Houston presentation, Williams dashed to his hotel room. He quickly returned to the hall and soon was telling commissioners about the difficulties in the oil patch. Then he started crying and buried his face in a hotel towel, which he called industry's crying towel.

The performance was a hit.

The 1986 TRC state of the industry hearing was held in Austin. When Williams was called to make his presentation, he appeared wearing a post Civil War era cavalry uniform, his hat full of arrows. He told commissioners things had gotten so bad that he had been forced to diversify into the telephone business.

Momma wouldn't even lend him money, he said.

Dallas was the site of the 1987 hearing. By then, the Williams presentation had become a tradition. The audience anticipated his appearance.

And Williams didn't disappoint. He showed up wearing an eye patch, a leg cast, and an arm sling. He told commissioners it took true grit to make it through the price collapse of 1986, but he had survived and was healing. He then ripped off the patch, cast, and sling.

But Williams topped all previous performances at this year's hearing in Lubbock. A sudden silence that filled the hall when his name was called was shattered by the sudden scream of an ambulance siren.

Two men came rushing into the hall, one wearing a sheet with a sign upon which "OPEC" was scribbled, the other a banker's suit with a sign reading "Big East Utilities." On a stretcher they carried lay Williams, wearing only a hospital robe, cowboy hat, and boots.

When Williams was dumped at the feet of the TRC commissioners, he launched into a tirade about outside influences beyond his control. With that, two women in nursing garb helped Williams climb into a wheel chair, which was no easy task considering the brevity of the hospital gown and the ungainly cowboy boots.

Williams called everyone's attention to signs identifying the nurses as Miss Efficiency and Miss Smart Drilling. These, he said, were his survival strategies.

What's in store for 1989?

Williams, who recently visited OGJ's Houston offices, says he's looking for suggestions.

Environmental big business

Bob Williams **November 21, 1988**

Environment. Big business. The terms usually appear in the context of opposition, as in "Environmental groups oppose development by big business in wilderness areas" or "Big business opposes new environmental regulatory roadblocks."

As for industry and environmentalism, big business means Big Oil — the

Seven Sisters plundering Mother Earth. Strapped independents with a few strippers are never hit by environmental costs, right?

On the other hand, environmentalists are altruistic souls in blue jeans, save-the-whale t-shirts, and sandals (all natural fibers, of course) ready to put their lives on the line to save sloe-eyed baby animals from extinction by voracious Big Business. An environmentalist could never be some $200,000/year Harvard lawyer, right?

Or is environmentalism itself getting to be pretty big business?

In California, where worship of the environment is outstripped only by worship of money, automobiles, and real estate, you can strike it rich by turning in violators of Proposition 65. Turn in anyone discharging near drinking water sources a substance suspected of causing cancer or reproductive damage and you collect 25% of any fines assessed. At a fine ceiling of $2,500/day, that could add up to a lot more income than your average stripper well could provide.

W. R. Grace & Co. Chairman J. Peter Grace, citing jury awards of hundreds of millions of dollars, calls for reform of toxic tort laws before "they ruin the private sector."

"We are forced to pass on the cost of these outrageous judgments to our customers as a shared liability," Grace said. "But sharing the liability for this sort of ridiculousness is . . . tantamount to highway robbery or getting mugged."

Environmental service is getting to be pretty big business, too. ENSR Corp. earlier this year predicted the environmental services industry in the U.S. will top $5 billion by 1990. ENSR noted that EPA will add more than 400 regulatory "products" to the market in 1988.

In addition to proceeds from the sale of everything from calendars to coffee mugs and regularly skimming cream from industry coffers under "charitable contributions," there are big entrepreneurial opportunities for environmentalists involving government.

Heritage Foundation has taken issue with a bill by Rep. Mo Udall (D-Ariz.) designed to create a trust fund using revenues from offshore drilling to finance added public land purchases by federal, state, and local governments.

Heritage's Kent Jeffreys, calling the bill "a billion-dollar boondoggle," notes that it also would make private, nonprofit organizations eligible for the same grants as long as they don't use the funds for administrative expenses.

"One wonders if administrative expenses would include those Sierra Club promotions bragging about the $45,000 annual income pulled down by the average member's family," Jeffreys said. "A group like that is obviously in need of federal assistance."

That's food for thought for those wondering if they're in the wrong business while oil prices are nosediving.

Conserving gasoline

Jim West **December 5, 1988**

Anyone who has been around very long knows the way the petroleum industry's trends come and go, many of them fueled by government policies.

In the late 1970s the accent was on all sorts of fuels as alternatives to oil. That was because President Jimmy Carter and Energy Sec. James Schlesinger were convinced there was no more oil to be found in the U.S. They didn't want too much demand chasing too little oil, whose price was rising.

So, the reasoning went, it was imperative to push development of things like tight gas sands, oil shale, and solar and wind — even tidal — power. And if by chance someone managed to keep producing oil a special tax would prevent a ripoff of consumers. It probably would be a good idea to wrap it in the guise of a "windfall profits" tax instead of calling it what it was, an excise tax that had nothing to do with profits.

Today, after that tax raised about $77 billion in its 8-year life, the accent again is swinging to alternative fuels but in a much narrower range and a much less dramatic way.

U.S. government policy today seeks to boost substitutes for gasoline. That policy is embodied in the Alternative Motor Fuels Act signed into law Oct. 14 by President Reagan. The law provides incentives for manufacturers that produce cars able to run on ethanol, methanol, or compressed natural gas.

The main goal of the law is cleaner air, although some industry spokesmen point out that a reduction in oil imports will be an added benefit. Warnings abound that the U.S. is becoming too dependent on imports.

In essence, about 10 years after the Carter-Schlesinger era, we have an about-turn in circumstances. There is too much oil chasing too little demand, which is rising.

Those interested in reducing air pollution and stretching supplies of a finite resource through substitution or conservation might take a page out of the University of Waterloo's book.

Last summer a team from the university, in Oakville, Ont., won the Shell Canada Fuelathon for the second year in a row, achieving 3,481.9 mpg in its vehicle "Astral." Propelled by a 70 cc Honda engine, it weighs 80 lb and is built of aluminum covered with plastic.

Quite a mileage performance. Right?

Not really. The UW team failed to beat its record of last year: an astounding 4,193.8 mpg.

Imagine. This year's performance of nearly 3,500 mpg didn't beat the record. Now that's conservation.

Shell Canada conducts its Fuelathon every year to encourage fuel economy research. Engineering students from Canada and the U.S. design, build, and drive fuel efficient vehicles around a set course to see who can go the farthest on the least amount of fuel.

Top student entries receive $3,500 in prize money from Shell Canada and $500 from the Ontario Ministry of Energy.

Second place in this year's event went to Shell Canada's Oakville research Center in its "Litre Beater" vehicle.

Role of futures expands

John L. Kennedy **January 23, 1989**

Not every firm uses energy futures. But each year, more companies participate in the futures market to help minimize price risk.

Many still would prefer to buy and sell in a stable market, where postings might not change for months. But there is no such market.

It is an understatement to say that the industry was reluctant to enter the futures market when heating oil futures were launched on the New York Mercantile Exchange 10 years ago last fall.

Rosemary McFadden, Nymex president, says Joel Faber, one of the original heating oil floor brokers, recalled at a breakfast marking the anniversary that traders sometimes waited over an hour for orders to come to the ring from outside. Daily volume might total 40-50 contracts.

Faber also told of his difficulty in subscribing to Oil & Gas Journal then. He borrowed it from a fellow trader who could subscribe because of other oil business interests.

That's how enthusiastic the industry was for energy futures.

When the price of crude began its roller coaster ride in 1986, industry's enthusiasm — if not for energy futures, for survival — started to pick up. When it became apparent that price volatility would last, the role of energy futures snowballed. The Nymex WTI contract is the most prominent example.

Currently, says Phibro's Andrew Hall in the special Petroleum Futures Report beginning on p. 33, ". . . the price paid for the majority of the almost 50 million b/d of crude oil consumed in the free world is determined by the price bid or asked for maybe 1/100 of that volume of 'domestic sweet crude' at Cushing, Okla."

London's International Petroleum Exchange also had its best year yet in 1988. Almost 1.7 million futures contracts were traded on IPE between January and November, vs. 1.1 million for the year before.

As of now, there is considerable reluctance by gas producers to embrace the futures market. But the contract set for launch on the Nymex this year may well follow a track similar to that followed by crude on its way to trading a volume of crude oil futures contracts that last year roughly represented the free world's oil consumption.

There is already an active spot market in natural gas, and prices can be highly volatile. Competition among gas sellers increases steadily.

Energy futures may not be the way oil and gas producers, refiners, or marketers would like to buy and sell crude, natural gas, or products. They still may not all be entirely comfortable with the concept. They still may not all be convinced that futures markets don't cause volatility, they only reflect it.

But for a growing number of companies, none of that matters.

For now, at least, they must protect the value of inventory or production, or guard against dramatic increases in feedstock or products costs.

Part of their strategy last year was to trade 33 million energy futures and options contracts on the Nymex, and nearly 2 million contracts on the IPE.

Santa Barbara revisited

Gene T. Kinney **January 30, 1989**

We are indebted to a faithful reader, William E. (Bill) Kennett, for the idea of taking a fresh look at the Santa Barbara Channel oil spill 20 years later.

As exploration manager of a major company operating off California at the time, he had a ringside seat to that landmark event and the rage that followed. No other force so shaped the industry atmosphere in the last two decades.

Jan. 28, 1969, was, to Kennett, a day of infamy in the history of the industry. "It was a shot heard around the world; it moved the environment movement from the back seat into the driver's seat. The oil industry has never fully recovered from the oil spill of that day."

Indeed, the industry hasn't been the same since. Offshore development — in fact, almost all oil industry activity, including issues totally unrelated to the environment — became even more politicized. The fallout triggered lease sale cancellations, a revolutionary National Environmental Policy Act, delays in constructing all manner of important facilities, including the Trans-Alaska pipeline, which finally took an act of Congress to complete. It even played a role in repeal of percentage depletion for major companies, all of which somehow had to be punished for this single accident. The spill inspired regulations and lawsuits that made it difficult to drill off the coasts of friendly states in the Gulf of Mexico, almost impossible anywhere else offshore.

Now, 20 years later, Kennett recalls the dire pronouncements made after the spill. Tourism was said to be finished. The Santa Barbara Channel would become a marine desert. "The facts," he said, "were lost in a forest of extreme emotionalism and false journalism."

The spill certainly made an awful mess. It fouled the channel and the beaches beloved by outdoor-minded Californians, taking a toll of marine life in the process. TV pictures of oil-soaked gulls, mired helplessly in the goo, were etched forever into public consciousness.

But terrible though it was, such messes can be cleaned up. Through a monumental effort that corner of paradise was quickly restored to a pristine state. The beaches, in fact, were made cleaner than before. Marine life recovered, too, and now prospers in harmony with oil operations.

Santa Barbara today is very much alive and healthier than ever.

"Tourism has increased many fold to the point where some citizens are doing all they can to slow it down," observes Kennett, a Santa Barbaran still active as a geologist. "You should see the improvements on the wharf and along Cabrillo Boulevard, the principal waterfront drive. Some marine desert!"

Would that the same could be said of the other mess.

Santa Barbara was a godsend to that sector of environmentalism that can't bear the temporary intrusion of a rig on the landscape.

But it was a catastrophe for the industry and the country. The nation will keep paying the price unless it learns to separate real environmental needs from the rest of the movement's agenda. Until a new balance is struck, the spill will continue to be far more damaging to the U.S. and its oil consumers than it ever was to Santa Barbara itself.

Communicator's goal

Jim West **February 6, 1989**

Any editor worth his salt has a collection of pet peeve words and mystifying phrases squirreled away to remind him that what crosses his desk every day, including his own work, is the product of human frailty.

The collection amounts to a cache of imperfect stories produced by imperfect professional writers and others, all trying to convey information. Sometimes they succeed. Sometimes they fail.

Among the worst offenders in playing fast and loose with the language are corporate communicators, especially those who prefer buzzwords to impress rather than simple English to inform.

Proof of that lies in press releases like those that tell of creation of posts such as manager of environmental affairs because a company wants to become more "proactive" in such matters. Proactive does not appear in any of the dictionaries in this office.

Or how about a warning from a Detroit auto maker that without relief from federal mileage standards it will have to resort to "foreign outsourcing" for certain goods? Some people call that importing.

There is a danger that in trying to keep things simple one can get too simple. That happened when one writer tried to make it clear that certain equipment had to be ordered in pairs of two. Another placed a footnote on a table to explain that the percent change between two columns was the percent difference. A doctor writing a column for an industrial newsletter revealed that "physiologically, women are the primary childbearers." And a brochure for a conference on diving and related technology cautioned the foolhardy, "Underwater work is impossible without air."

A commissioner of revenue for a certain northern U.S. state, probably aware of government's penchant for fuzzy writing, said of an oil revenue forecast: "We believe the report is unusual in several ways. The report is at all points clear and direct."

Resisting a possible attempt to be pinned down, another writer's press release said his company's distribution to limited partners will amount to "approximately $0.07693/unit."

That's clear enough — certainly clearer than some translations of press dispatches from other languages into English. The problem is that certain words have no equal meaning in other languages.

That appeared to be the case when an Asian news agency reported, "Libya will export crude oil to Morocco, which in turn will export edible oil and gas cylinders to Morocco." There were no details on what has to be world's strangest trade arrangement, so we didn't run the item.

Nor did we run a story based on a dispatch about another country's petroleum research center that included "a rig experiment station and underground petrochemical blast laboratory." Probably very few employees volunteer to work in that laboratory.

Of course, it's possible to be mystified by some words in English, the language of the international oil industry. That occurs when stories from our international editor, Roger Vielvoye in London, send us to the dictionary to find the meaning of words like wrong-footed, liverish, vetting, jiggery-pokery, in a cleft stick, and knock-on effect.

Like Roger, the rest of us at the Journal try to communicate clearly. Our goal is to succeed more than we fail.

$700 aspirins

Bob Williams **March 6, 1989**

Hi, folks. Planet Earth here, subbing as guest columnist this week.

And a pretty sick little orb at that, or so they tell me. I really wasn't feeling all that bad until suddenly last summer it was gloom and doom about yours truly. Makes one a tad nervous, especially when the state of your health hits Time and Newsweek covers, making it official.

It seems my worst maladies are this touch of acid rain and this tinge of greenhouse effect that's getting worse. Or maybe it's not. Beats me. A lot of experts are pointing at changing my diet of fossil fuels, although a lot of other experts say that's only a small part of the cause.

There are about a jillion "doctors" out there, each with his own cure. If you have been sick lately, you know that getting well isn't cheap. We've all heard the one about the $7 aspirin for the hospital patient. Sure, a case could be made about the hospital overhead, expensive tests, and the fact that the aspirin couldn't hurt and probably helped.

In my case, the "aspirin" is more like $700, it may not be needed, and it could trigger an allergic reaction. I can hear it now: Whoops, Mr. Earth, those estimates were wrong. The aspirin cost $700, and you have a new and worse medical condition as a result. Guess we should have waited for the test results to come back from the lab.

Wouldn't you be a little ticked?

Like A. Joseph Dowd, senior vice-president of American Electric Power (AEP) says, Congress should at least see the results of the most comprehensive acid rain study undertaken — the 10 year, $500 million National Acid Precipitation Assessment Program (Napap) due for completion in mid-1990 — before passing new acid rain laws.

"It is just plain common sense for Congress to wait another year and a half until the results of the Napap study are in," he said, adding that "to do otherwise would verge on the irresponsible."

Whoa! Congress and common sense? What is this guy, a radical?

Dowd said the last acid rain proposal in Congress would have cost AEP's power system $1.3 billion by 1994 to retrofit older plants with scrubbers, forcing electric rate hikes of as much as 21% for residents and 48% for industries.

At least there's certainty among experts over the greenhouse effect, right? It's getting warmer. Or colder. Or it's staying the same. And it's caused by burning too much gasoline and coal. Or tropical deforestation. Or methane from rice paddies, termites, and cow flatulence. And some say any effort without some of my Third World parts helping would be futile.

Stephen Schneider of the National Center for Atmospheric Research admits experts agreeing on the existence of a greenhouse effect "are looking into a very dirty crystal ball."

He says, "The tough problem is how long we clean the glass before we act on what we think we see inside."

Schneider favors immediate environmental action as an insurance policy against "the uncomfortable risks of future grim surprises."

Wait a sec. I don't feel terminal yet. I wouldn't mind a little more crystal ball polishing.

I just worry that, one way or another, all those $700 aspirins are going to kill me.

Chemical engineering 101

Bob Williams **April 3, 1989**

The more things change . . .

In 1989, year 1 of chemical engineering's second century, a major concern for U.S. refiner/marketers will be how to meet soaring gasoline demand when domestic supplies may be constrained.

That was essentially the scenario in 1915, when chemical engineering had its first true practical application on a commercial scale, says a National Research Council study on the future of chemical engineering. University of Houston Senior Vice Pres. Neal R. Amundson spearheaded the study.

Chemical engineering was born in 1888 when Prof. Lewis Mills Norton started a chemical engineering program as an option in the Massachusetts Institute of Technology chemistry department. The curriculum mainly described industrial operations, organized by product, and the lack of an intellectual paradigm soon became apparent, the Amundson report said.

The discipline's first paradigm was based on the unifying concept of unit operations. Proposed by Arthur D. Little in 1915 in response to the need for commercial scale manufacture of commodities, it held that any chemical manufacturing process could be resolved into a coordinated series of operations.

That idea was put to its first large-scale test 5 years later, in time for the first U.S. gasoline crisis.

"The rapidly growing number of automobiles was severely straining the production capacity for naturally occurring gasoline," the Amundson report said. "The ability of chemical engineers to quantitatively characterize unit operations such as distillation allowed for the rational design of the first modern oil refineries. The first boom of employment of chemical engineers in the oil industry was on."

Chemical engineering and the petroleum industry have intertwined in other U.S. crises. Chemical engineers were crucial in helping develop sources of synthetic rubber to replace natural rubber supplies lost to the Japanese early in World War II. Chemical engineers at Shell Development Co. designed a pilot plant to boost recovery of synthetic penicillin to about 85% from about 33%. The nearly sixty fold hike in U.S. penicillin production when the Shell process was installed by four producers saved the lives of untold thousands of soldiers and civilians.

The second paradigm for chemical engineering was inspired by the engineering sciences movement in the U.S., says the Amundson report. Dissatisfied with empirical descriptions of process equipment performance, chemical engineers studied unit operations at the molecular level and developed quantitative models critical to the birth of the U.S. petrochemical industry — a business again on the rise.

The Amundson report sees a third paradigm emerging in response to, among other things, an explosion of new products and materials, increased competition in world markets for commodity chemicals and petroleum products, and environmental, safety, and health concerns.

A tale of two crises

Rick Hagar **April 10, 1989**

The residents of Norco, La., were jarred awake shortly after 3:30 a.m. last May 5 by a Shell Oil Co. refinery explosion that rocked the entire city. Within minutes, sleepy Shell media relations people across the U.S. began receiving telephone calls from news reporters wanting details.

About 10 months later, Exxon media specialists were awakened by some of the same reporters. This time, it was history's largest oil spill in U.S. waters, which occurred when the Exxon Valdez supertanker ran aground in Prince William Sound off Alaska shortly before 1 a.m. Mar. 24.

Industry can learn some lessons from how Shell and Exxon dealt with the media during each company's crisis.

In Norco, reporters that had gathered outside the refinery gates were escorted to a nearby administration building, where the plant's community relations staff began the first of several news briefings.

The reporters had an immediate problem, however, because they had access only to the pay phone at a convenience store across the street. Shell moved quickly to install a phone bank in the news room, which allowed all reporters to meet their morning deadlines.

At about the same time, the staff began sending bulletins to Shell offices across the U.S. Reporters unable to get to Norco in person or by telephone could call the nearest Shell media representative for an update.

Corporate headquarters in Houston fielded questions about the business implications of the explosion. Shell did all it could to communicate with the media.

Exxon also set up a media center. But it was in Valdez, where the infrastructure was incapable of handling the flood of reporters and their telephone calls. Exxon did not update its media relations people elsewhere in the world. Houston, New York, and Los Angeles told reporters it was Valdez or nothing.

Exxon issued no immediate public statement, while the National Audubon Society sent reporters the names and telephone numbers of a wildlife biologist, an ornithologist, and two policy analysts. There were no Exxon press releases, but the Sierra Club distributed a 580-word statement from its chairman.

Exxon Chairman Lawrence Rawl's first public statement came Mar. 30, and then only to Reuters U.S. Inc. and the MacNeil/Lehrer NewsHour.

It should be no one's surprise that early worldwide news coverage of the spill mostly quoted government officials, angry Alaskan fishermen, and outraged environmentalists.

C. J. (Pete) Silas, chairman of Phillips Petroleum Co., told a Houston audience Mar. 29 that the media are partly responsible for creation of an aversion among U.S. businessmen to risk taking because too many times only one side of a story is told.

The point is well taken, and is one upon which several of us base our success. But businessmen need to understand that unless they are willing to make readily available their side of the story on a timely basis, only one side can be told.

Apples and oil

Bob Tippee **April 17, 1989**

While the Exxon Valdez was spilling crude oil into Prince William Sound last month, apples were making a comeback in U.S. schools.

They had been banned due to concern that the growth regulator daminozide — known by the trade name Alar — gives kids cancer.

To be precise, it was more than concern. It was fear.

In February, the Natural Resources Defense Council warned that children face high cancer risks from chemicals used on food products, especially Alar. The Environmental Protection Agency announced it would phase in a prohibition against Alar due to laboratory indications that the substance might threaten humans, especially kids. And actress Meryl Streep starred in a television commercial in which she scrubbed broccoli with detergent and fretted — convincingly, as always — about how Americans poison their offspring with pesticides on farm produce.

Scary stuff. School systems in New York, Los Angeles, Chicago, San Francisco, Atlanta, and elsewhere made apples verboten. Some of their youthful charges undoubtedly resorted to chocolate-frosted sponge cake and other such healthful delights. Then science took over.

EPA and two other agencies denounced NRDC's risk analysis. Apples, they declared, pose no hazard to public health, treated with Alar or otherwise. Yes, EPA will phase out Alar, which is disappearing anyway. But the move is precautionary, not an emergency response to suddenly discovered poison in human food.

So apples came back. And Exxon Valdez ran aground, providing new fodder for this strange apparent need of Americans to be terrorized by their surroundings.

Human beings are weeping for dead birds and sea otters. Some are declaring all of Prince William Sound dead. Afraid of a recurrence, politicians are calling for a halt to offshore leasing and drilling.

Tragic as the spill is, some of these responses are as appropriate as jerking apples out of school cafeterias.

A sea captain with a problem made an error that killed animals, created a dreadful mess, and disrupted human lives and livelihoods. But the cleanup has begun. Those who suffered harm will be compensated. Frightened people, nevertheless, are responding as the fearful will — with outrage and vengeance, neither of which will clean up Prince William Sound.

Just as evidence of carcinogenicity in laboratory rats differs greatly from — if it relates at all to — mass poisoning, spilled oil soils but does not kill a waterway. The oil, not birds and otters, will go away.

But fear smothers distinctions like that. Outrage and vengeance hinder solutions.

For a chastened oil industry, fear will fuel political antagonism for years. It will thrive on sudsy vegetables in the hands of beautiful celebrities, on oil-soaked bird carcasses in the hands of mourning environmentalists. It won't go away, but it can and must be resisted.

That's the long term challenge. Short term, industry should examine immediate causes and effects. If prohibitions are in order, they should address the real problem in the Exxon Valdez disaster: Don't give employees the chance to play environmental roulette with alcohol.

Play it safe. Give them apple juice instead.

Natural gas — free at last

Gene T. Kinney **June 5, 1989**

Events like the Phillips decision, a symbol of judicial lawmaking at its worst, aren't the kind usually celebrated by the U.S. gas industry.

But this year's observance on June 7, the 35th anniversary of the Supreme Court ruling extending federal price controls to the wellhead, should be different. If all goes well, which itself would be a break with historical precedent, the great wrong visited upon the industry and consumers is about to be righted — again.

Congress will shortly return to producers the freedom it has tried to grant them on three other occasions during the past 50 years. The fourth time may be the charm. If so, it will end a long battle marked by frustrations that are surely unique to this issue.

When Congress enacted the regulatory scheme for interstate gas pipelines in 1938, it expressly exempted production and gathering. But the treatment of producer sales was controversial almost from the beginning. The Federal Power Commission, charged with implementing the Natural Gas Act, leaned first one way, then the other. The question was taken up in the Phillips Petroleum Co. case, started by the commission in October 1948. Sen. Robert S. Kerr (D-Okla.), moving to head off regulators at the pass, rammed through the Kerr gas bill in 1950, expressly exempting producer sales to interstate pipelines.

However, a campaign by muckraking columnist Drew Pearson generated a flood of mail to the White House — more heat, President Truman wrote in his memoirs, than he took on any issue except the firing of Gen. Douglas MacArthur. His veto resulted.

When FPC completed its Phillips investigation in 1951, it reaffirmed that producer sales were not subject to regulation.

But an appeal by Wisconsin, whose counsel didn't expect to win and decided to go ahead almost on a coin toss, changed all that. An activist Supreme Court shocked everyone by siding with the state.

In 1956, in the plainest language yet, Congress passed a third deregulation bill. Once again, it was thwarted at the White House. This time it was an 11th hour campaign contribution from an oil company lobbyist that inspired a veto. President Eisenhower called for new legislation with similar provisions. But it was not to be.

Since then, regulators have struggled in vain to find a formula that worked. A divided industry has complicated efforts at congressional relief, and the Natural Gas Policy Act of 1978, the first producer regulation legislation ever enacted, only made matters worse.

Now, with supplies declining and the ultimate absurdity of regulated prices that are above the free market, the bankruptcy of regulation is evident to all. Deregulation has become motherhood — in the Federal Energy Regulatory Commission, the White House, and Congress. The House of Representatives has taken the initiative by passing a bill. And the Senate should follow this summer. The industry has but to accept it.

After ignoring producers for the last two decades, the world is finally getting the message of the marketplace.

So, barring yet another jinx, it appears appropriate for the first time to say, "Happy anniversary, Phillips."

Elephants and oil

Bob Tippee **July 17, 1989**

How fitting that it all should have started with a bossy Englishman and an untidy elephant. . . .

India's oil industry, dated from discovery of Digboi field in northern Assam, turns 100 years old this year. It has come a long way in a century. It now produces about 650,000 b/d of oil and 860 MMcfd of natural gas. Reserves at yearend 1988 totaled an estimated 6.3 billion bbl of oil and 22.8 tcf of gas.

But it's not enough for a country with 2.4% of the world's land area, 15% of its population, and consequent economic strain. India must supplement domestic production with more than 360,000 b/d of imported crude to feed its 12 refineries. And it imports 6-7 million metric tons/year of products. Demand, growing at the rate of 7%/year, will reach 2 million b/d by 2000.

So the oil industry hustles. Two state owned exploration-production companies, Oil & Natural Gas Commission and the smaller Oil India Ltd. (OIL), plus Soviet crews operating under an exploration protocol, kept 137 rigs busy during May. They'll soon be joined by international companies drilling on eight offshore production sharing contract areas awarded in the third licensing round. The fourth round, now in planning, might include onshore tracts.

Exploratory results have been spotty, with nothing rivaling the 1974 Bombay High discovery off the west coast. Bombay High and three nearby fields account for nearly 70% of India's oil production and 80% of its gas output. Development continues in those fields and in recent area discoveries.

But promising discoveries have turned up elsewhere, especially in Gujarat, in Cauvery basin in the Southeast, and in the Himalayan forelands of the North.

To make use of growing gas reserves, India hopes to develop a national grid anchored by the 1,094 mile Hazira-Bijaipur-Jagdishpur pipeline completed last year. And it plans to keep crude oil distillation capacity within

85% of total demand by building refineries at Assam, Mangalore, Karnal, and three other sites yet to be determined.

India's petroleum industry, in other words, is busy. But it won't let its centennial pass without celebration.

On Nov. 4, officials will release a postage stamp and book commemorating oil's hundred year milestone in India. They'll make Digboi field, which still produces, a national park. The next day they'll conduct a seminar and exhibition in Duliajan, Assam. And on Jan. 24, 1990, OIL, Digboi operator and centennial coordinator, will conduct a petroleum technology seminar at Vigyan Bhawan, New Delhi.

Quite a program, and quite a legacy, for something that began with that imperious Briton and his messy pachyderm.

As OIL tells it, an elephant carrying logs through the jungle for Assam Railways & Trading Co. in September 1889 entered camp with oil on its feet. Workers tracked the animal back to a seep, whereupon an excited English overseer issued the fateful order, "Dig, boy, dig!" And India's first oil field got its name.

Mothers of invention

John L. Kennedy **August 21, 1989**

Twenty years ago next month, the Journal reported that Phillips Petroleum Co. had a big oil strike in the Norwegian sector of the North Sea. Phillips wasn't ready to talk about it yet, but other sources said the field was being named Echofish.

Six weeks later, reports were circulated that the discovery had been confirmed by a second well.

At the time, the score by nine operators in Norwegian waters was 29 dry holes, three producers (Phillips), and 1 suspension (Phillips).

Not encouraging — at least not to eight of the nine operators.

But by early spring, Phillips felt it had opened "a new petroleum province" in the North Sea with its first Ekofisk well and drilled a step-out. Then in early June 1970, Phillips made it official, saying, "It appears we have discovered a giant oil field."

Oil finds proliferated after the Ekofisk discovery. And technical innovations came even faster.

Many of the highly-publicized ones were achieved at Ekofisk — improvements in directional drilling techniques, the concrete storage tank, ultrahigh pressure compressors, the jackup operation in 1988 to offset seabed subsidence, this year's installation of the concrete barrier around the Ekofisk tank.

But hundreds — no, thousands — of large and small innovations in design, in operation, and in maintenance have been absolutely necessary to exploit North Sea reserves.

The first mothers of invention were a harsh environment and high costs. New structural designs and temporary producing facilities were just two of the results.

Economics has been the latest impetus for innovation.

In 1969, the posted price of Saudi light was $1.80/bbl. But while the North Sea's early oil producing and transportation system was being built, prices got better and better.

The main concern was technical: was it strong enough, did it have enough capacity?

When that all changed in 1986, more efficient designs, more measured development, more attention to trimming excess operating costs brought another round of innovation.

Our special report on Offshore Northern Europe, in this issue, shows the early results: renewed exploration and development activity and a promising future.

It also cites the latest — maybe the most serious yet — challenge to North Sea innovators: safety.

Response to this challenge has already begun. Changes are being made in safety systems, procedures are being reevaluated and revised, and new regulations are in effect. More changes in facilities design criteria and operating procedures are certain.

If the record is any indication, another wave of innovation will improve North Sea safety.

G'day, mate

Rick Hagar **August 28, 1989**

The northern region of Australia, known as the outback home of movie character Crocodile Dundee, also is home to one of the world's hottest exploration/development plays.

While Dundee takes fictional walkabouts, exploration companies punch real holes in the Timor Sea, the stretch of water that separates northern Australia from Indonesia.

Timor Sea operations are among the most isolated in the world. That fact is highlighted by BHP Petroleum Pty. Ltd.'s extensive operations to simply transport crews to and from Jabiru Venture, the converted tanker used to process and store crude from the region's only producing field.

Red Beans, Rice, and Tabouli

Headquarters for Jabiru Venture operations is 400 miles due east of the field in Darwin, the Northern Territory frontier capital where petroleum engineers and roustabouts hoist Green cans and dine on barramundi with ringers and jackaroos from inland cattle stations.

To avoid the rigors and hazards of direct flights between Darwin and Jabiru Venture, BHP sends its crews first in fixed wing aircraft to a jumping off point on Troughton Island. Although about 335 miles southwest of Darwin, Troughton Island is the nearest coastal island to Jabiru Venture large enough to support a runway.

First-time passengers on the trip are surprised to see a flight attendant distribute magazines and serve soft drinks and food. BHP explains that such creature comforts are needed to relieve the monotony of the 90 min first leg flight and as an incentive to attract good crews to the frontier life in Darwin.

The next surprise is Troughton Island itself, which originally was the site of a shipping lane radio beacon. It is only 3,000 ft long, 1,500 ft wide, and 13 ft above sea level at its highest point.

A compound in the center of the island is staffed by 19 men and women. Two seawater desalination units provide water, mixed with rainwater for taste.

In the center of the compound is a swimming pool, which seems strange at first considering that the beach is only a few feet away. The explanation is chilling: The Timor Sea is inhabited by an assortment of sharks, saltwater crocodiles, and sea snakes.

After landing on the island's dirt runway, passengers get a snack in the compound's mess hall in preparation for the final leg of the trip, a 150 mile ride by helicopter to Jabiru.

The 60 min flight passes quickly as passengers contemplate the loneliness of Timor Sea operations. The only other sign of life spotted is a supply boat heading to port from Jabiru Venture.

At such times, passengers tend to recall BHP's training facility in Darwin. It consists of a large metal box that is dropped into a deep pool of water and rotated. Configured as either plane or helicopter, it is used to teach BHP's crews how to safely get out of a craft that has ditched and flipped over.

Crocodile Dundee may get all the publicity with his Hollywood trips into the lonely outback, but the true Australian explorer today works in the Timor Sea.

Tomorrow's horsepower

Gene T. Kinney **October 16, 1989**

Can man provide the energy required to raise living standards in the next century in a form that will not destroy Planet Earth?

This question was tackled by dozens of scientists, economists, and government officials at the World Energy Conference, where more than 4,600 delegates from 88 countries discussed the answers last month.

Predictably, they started with a common premise: Future energy growth must be supplied in a less polluting manner than in the past, or sustainable energy development will not be possible. And all assigned a key role to technology.

The program flavor was so pronounced that one speaker described it as an environmental conference at which energy was discussed, rather than the reverse, and under sponsorship of the nuclear industry, which many advanced as a solution. The description was not far off.

Under assumptions of a new WEC forecast, the scientists and engineers have their work cut out for them. During the 35-year period starting in 1985, world energy demand is likely to grow 50-75%. In spite of conservation and pollution control in the industrialized countries, population and economic growth in developing countries will boost energy demand, especially for the fossil fuels and firewood that are the biggest polluters of the air.

In the aftermath of Chernobyl, "nonpolluting" nuclear energy will increase its market share to only 7-8% in 2020, well above the 4% in 1985 but far short of the 12% forecast by WEC in 1983.

Natural gas, deemed the major short-term beneficiary of heightened environmental concerns, is due to keep its 18% share, simply growing with the market, while oil will drop from 32% to 26-28%.

Surprisingly to many environmentalists, that leaves solid fuels to play an expanding role, to 28% or 30% in 2020, from 26% in 1985.

A major disappointment is the growing role of noncommercial sources, mainly firewood, with its threat of deforestation, and the limited growth of renewable sources. Austerity promotes the former while low oil prices and stagnant technology restrict the latter.

This outlook, reasonable though it seems today, may contain a fundamental flaw: underestimation of the impact of new technology. While inherently impossible to predict, surprises from science are still a certain part of the future. Only the degree of impact is in question.

To illustrate, one speaker cited a forecast of a 100 years ago. Future transport needs of the city of London were predicted accurately but were thought to be impossible to meet. The daily pollution from the horses required would have reached 3 m deep.

The opposite error was shown in another projection. This frivolously postulated the potential conversion of all matter into energy, leaving earthlings with no option but to move to another planet.

Frivolity aside, the real energy future lies somewhere in between. Science will again provide horsepower in a less-polluting form, though not necessarily in a vehicle recognizable today. And interplanetary travel will become an option for many, though hardly a necessity.

Live from Dilley, Tex.

Guntis Moritis **February 26, 1990**

Was it fire, flood, or some other catastrophe that was about to be covered by a Houston television channel live from a place named Dilley, Tex.?

No!

TV had discovered an oil boom in South Texas. Horizontal drilling was at the heart of the activity, and in the background local politicians seeking higher office proudly claimed credit for the event.

The boom is real. The drilling around Dilley in the Pearsall Austin chalk area will make money for landowners, speculators, operators, and service and equipment companies. But the total volume of oil produced will not dent the amount imported into the U.S.

Much more drilling is needed.

Although horizontal wells represent only about 1% of all wells drilled last year, the growth of horizontal drilling is exciting.

As Oil & Gas Journal's second annual survey on horizontal wells reveals, the future looks bright for this technology.

Horizontal drilling is catching on all over the world and in most instances achieving success. In most cases success came only after rigorous technical analysis. Not all areas are appropriate for this technology, and caution is necessary in selecting applications.

Saving money

Horizontal wells can produce more reserves in a shorter time for less money. Therefore, they can make more areas economically attractive. But in the U.S., unless a field is unitized, getting the optimum use from money needed to develop reserves is difficult at best.

The system for well placement in the U.S. is determined by how best to drain the largest possible area.

The drilling of a well by one operator may move offset operators to drill to avoid drainage. In this manner probably more wells are drilled than needed to produce a given reservoir.

Some oil may be lost by not being the first to drill, but the second to drill benefits because there is less risk of a dry hole in an offset well. In most cases, less risk is the important factor, and leases aren't drilled until near expiration or unless a discovery is made on an offset lease.

This system can create a speculation frenzy when significant discoveries are made.

A lot of money goes into boom areas that is not used directly to produce oil.

Haste can waste money and prevent optimized field development, but without this system the Austin chalk might still be lying fallow.

More orderly development, although it might bring better recovery in some cases, could risk stifling those who try innovative technologies, like horizontal drilling.

Exxon Valdez wounds starting to heal

Bob Williams **April 2, 1990**

Wounds from the Mar. 24, 1989, Exxon Valdez oil spill in Alaska's Prince William Sound have begun to heal.

Signs abound that the sound's ecosystem is recovering from the spill, aided greatly by last winter's storms.

The city of Valdez issued a press release appealing to the world's press to avoid repeating errors and myths in covering the spill anniversary. It noted that the loss of wildlife was a small fraction of existing populations in the sound and that most of the affected shoreline is remote and outside the area likely to be seen by seaborne tourists.

Ironically, one of the daily sightseeing cruise lines working in Prince William Sound reports a record number of advance bookings — because the spill has made the sound a household name.

Exxon Valdez Capt. Joseph Hazelwood was acquitted of the most serious charges against him and found guilty of a misdemeanor, negligent discharge of oil, which he is appealing.

Repairs on the Exxon Valdez supertanker are about 70% complete and the ill-starred vessel should be returned to service by July with a fresh start and possibly a new name in the Gulf of Mexico, Exxon hints.

Industry's image

The spill's damage to industry's image is proving tougher to heal. Last month's first anniversary of that grim event provided a practice volley for the environmental lobby, amassing its war chest of hyperbole and doomsaying for Earth Day 1990 Apr. 22.

Environmentalists celebrated the Exxon Valdez birthday by calling on President Bush to end the U.S. "addiction to oil," comparing oil use to a dependency on crack cocaine. The Alaska Coalition called on Washington to establish a memorial for the wildlife lost in the accident by declaring the Arctic National Wildlife Refuge a wilderness.

Despite spending about $2 billion for a cleanup effort that a top executive with a competitor said, "probably no other company could have had the resources to manage," Exxon still reels under the weight of more than 170 lawsuits, faces criminal trial and still-mounting cleanup costs, and is pressured increasingly by large institutional shareholders to restructure to show a stronger commitment to the environment. Exxon may not ever fully recover from its wounds — which some deem self-inflicted and deserving as the latest icon of corporate villainy. Sad to say, much of that ire is aimed at innocent Exxon employees, many of whom spent months far from home and worked 20-hr days to combat the spill's effects.

But industry still has a good chance to recover from its wounds. TV commercials and public relations campaigns touting environmental commitment won't suffice, though. In the decade of the environment, industry will be judged by deeds, not words.

BP's success

That's why BP America Inc. was able to keep the Feb. 7 oil spill off California from turning into a PR disaster, instead winning praise for a successful cleanup.

Chuck Webster, BP America manager of crisis management, said, "It's amazing the number of local residents that came up to our supervisors and thanked them for the job they were doing."

In southern California!

This month, BP's crisis team will return to Huntington Beach, Calif., to host a dinner for about 100 people from area communities.

"We'd like to go back and say, 'Thank you for your help,' and carry on the process of lessons learned," Webster said.

The healing begins.

Crawfish and chili

L.R. Aalund **April 23, 1990**

With Oil & Gas Journal's exclusive biennial compilation and analysis of enhanced oil recovery projects in this issue, we could easily devote this column to the tribulations and small victories our staff has had since October while collecting this mass of data.

"Our staff" means this writer, Guntis Moritis, and Kim Jordan. We also had the essential help of scores of people in many enlightened oil companies who filled out or updated our questionnaires. We want to thank all and apologize to a few for badgering them for a response.

But like cowboys and cowgirls after a long cattle drive, we are ready for a new perspective. We'll let the EOR report speak for itself and go on to fresh topics — crawfish and chili.

Last spring we wrote in this column about the return of a big crawfish boil to the Offshore Technology Conference, which is again coming to Houston in 2 weeks. The point of that column was that the crawfish boil from Cajun Louisiana and Lapland (that region around Port Arthur where Louisiana laps over into Texas) has become as much a part of the offshore industry as jack-up rigs. The absence of the really big crawfish boils in 1986, 1987, and 1988 was a sure indicator of hard times.

The crawfish boil returns

Last year Benton Baugh, enthusiastic inventor, American Society of Mechanical Engineers representative to the OTC, OGJ author, and University of Houston graduate, initiated the return of a big OTC crawfish boil. It was to be a signal that the industry was fighting back. Naturally, the venue would be the University of Houston campus, and naturally Baugh wanted some publicity in OGJ.

But this Cajun feast was to be different from the historic ones. They were free. For this renaissance event, though, you would need a ticket and the profit, if any, would go to the University of Houston College of Engineering.

It turned out to be a success, with the rain coming the night before and about 1,000 showing up to eat remarkable amounts of excellent crawfish on the greensward under the shady oaks of Lynn Eusan Park. The event produced more than $3,000 in profit for the college.

Naturally, the organizers have scheduled it again for the same place the afternoon of May 6.

It is listed this time in the official OTC program and may be on the way to becoming an OTC tradition.

Another event born during the bad times is also thriving and becoming a tradition. It's Harrisburg Inc.'s fifth annual chili cookoff and barbecue. This got started during the crash of 1986 when Harrisburg opened its cavernous shops and warehouse in Northwest Houston to other oil field companies who individually couldn't afford to hold a big OTC party. Their customers weren't any better off.

A real Texas party

The hillbilly music was loud, the suds flowed, the chili was hotter'n a fire cracker, and the barbecue smoke hung so thickly over the floor you could hardly recognize anyone over 6 foot tall. It was a real Texas oil field party and fun.

This year it's still free to the offshore crowd, but it has gotten bigger and fancier. At least the smoke can drift. It's going to be on May 4 from 11 a.m. to 8 p.m. at the Great Southwest Equestrian Center at Katy, Tex., with more than 40 companies cooking and feeding new and old friends.

They say if you are having trouble getting off for it call in sick or bring the boss.

John McCaslin retires

John L. Kennedy **May 21, 1990**

No one on our editorial staff has seen more of the industry's cycles than Exploration Editor John C. McCaslin. And those cycles often have been particularly extreme for John's primary readers: oil and gas explorationists.

John will retire May 31 after more than 38 years as a Journal editor. During that time he has built an unparalleled worldwide following of geologists, geophysicists, and others involved in oil and gas exploration.

He joined the Journal in 1951 as district editor, then served 10 years as geological editor.

Since 1965, he has been exploration editor of the Journal and for the past 17 years has served as editor-in-chief of the annual International Petroleum Encyclopedia.

Almost every week during his 25 years as exploration editor, John has written about exploration prospects, activity, and results in virtually every geologic province.

In each week's feature article, he's brought to our readers the continuing advance of exploration technology and the petroleum geology of the world — from the arctic to Guatemala to Antarctica, from the Williston basin to Zaire to Papua New Guinea.

From the beginning, he has reflected his readers' enthusiasm for trying untested areas and the general optimism they bring to the worldwide search for petroleum.

Especially in the U.S., he's been on the ground in many of the major oil and gas provinces.

We'll miss the experience and knowledge that John has used to make the Journal a leader in providing exploration news and technology.

John's successor

Fortunately, John's successor will bring his own brand of experience — and equal enthusiasm — to the Journal's exploration pages.

Alan Petzet, who will become Journal exploration editor on June 1, is also widely known in the industry as a specialist in exploration and drilling activity and technology.

He's been with the Journal 9 years and has been writing about oil and gas operations for more than 20 years.

Alan, too, likes to see operations firsthand. Even on a weekend, he's likely to show up on a drilling location several hundred miles from the office.

Never more important

As of today, of course, the world has more oil and gas than it needs. But that's hardly relevant. Tomorrow, we'll need more; petroleum will have to get us through to the next family of energy sources as economic rules continue to change.

And even today, individual countries and companies don't have all the oil or gas they need.

All this means that exploration success has never been more important.

Our goal is unchanged as we make this staff transition. In more than 5 pages every week, we'll continue to bring our readers the important exploration news, geology, and technology.

Take your tone and . . .

L.R. Aalund **May 28, 1990**

Cave-like drawings under silent freeway overpasses and legends of isolated tribes that may inhabit this planet in the distant future may indicate that the breakdown in late 20th century civilization and commerce began with the appearance of the telephone recorded message.

Until then print and electronic communications were taking the world forward in giant leaps. Now progress is threatened, and the threat starts with "At the sound of the tone . . ."

Of course, like atomic energy, jet engines, and dynamite, the telephone recorded message can also be a force for good and productivity. But in the hands of the careless, ignorant, or discourteous, the recorded message is to communications what a millstone is to a neck.

A saving grace is that many recorded messages are so comical, whether coming from a major oil company, an airline, or your dentist's office, they are good for a healthy, stress-relieving laugh. The tonal quality of some must rival that of Alexander Graham Bell's first call. On the other hand, some are really good. This editor has been trapped into saying, "Hello, how are you?" to recordings.

Some experiences

What's our interest in recorded messages? Oil & Gas Journal editors, like most journalists, spend many hours trying to reach a variety of organizations and people. We are regularly exposed to the full gamut of the latest recording technology and systems.

Each has had a few distinctive experiences. This editor recently had a 2 month old call returned from a West Coast oil company researcher who was originally, according to the tape, simply "out of the office." One wonders how long a tape he has on his machine.

Then there was the message of a consultant in Washington that said, "I am in the office today but away from my desk."

That sort of courteous message is a major step forward in righting what is wrong with recorded messages. After all, isn't not knowing whether the person is just down the hall or on assignment in Shanghai one of the most frustrating aspects of a message?

He returned the call 2 days later from New Orleans.

That brings up messages with a lot of words but no information.

For example, dial a manager at one major oil company's operation in Texas and a smooth recorded voice tells you your call is being handled by such and such a system. After that commercial, the recorded voice of your target's secretary tells you he or she is, take your pick, "on the other line" or "out of the office."

But the secretary's voice goes on implying that if you are such a desperate creature you can get a live person by punching some more numbers. Ha! Gotcha! They left the recorder on, too. True story.

Then there is the one where you are shocked to get a human switchboard operator who, when you stutteringly ask for your party, gives you a recorded message that is being synthesized even as you listen.

Living with noise

An expert on noise pollution at a petrochemical plant in West Germany once told this editor people can to a degree live with identifiable noise like trash cans being slammed around. But it is the unidentifiable or "informationless" noise that drives people batty. The informationless recorded message does that.

Super gas eliminator

Richard A. Corbett **June 18, 1990**

Clean air concerns in the U.S. are spurring vehicle developments that take advantage of the emissions-reducing benefits of alternate fuels.

Designs for buses, trucks, cars, and other vehicles fueled by methanol, LPG, and compressed natural gas (CNG) are appearing throughout the U.S.

Many of these designs are test beds to prove the capability of vehicles and fuels because the public is not yet convinced that alternate fuels or alternate fuel vehicles work.

They want proof before buying either.

All of the developments aim to take advantage of emission-reducing fuels that are likely to be required during the 1990s and into the next century, as OGJ's special report, "Fuels for tomorrow," reveals.

CNG quarter-miler

Even in drag racing, an activity rarely noted for its environmental awareness, an alternate fuel racer is scoring notable wins.

CNG fuels the "Natural Gasser," a blue 1987 Buick Somerset dragster, owned and driven by Joe Mezquita of North Canton, Ohio. Joe, by the way, is supervisor of production for East Ohio Gas Co., Cleveland, supplier of fuel for the car.

Designed and built by Jeg's High Performance Center in Columbus, the car is a fiberglass replica of Buick's street version, mounted on a specially designed, drag racing chassis.

The dragster competes nationwide in the National Hot Rod Association (NHRA) Super-Gas class. That class of competition allows cars that use gasoline, methanol, and other fuels.

Powered by a naturally aspirated, 468 cu in. displacement Chevrolet V8 modified for CNG, the car puts out 700 bhp at 7,600 rpm. Engine compression ratio is 14:1.

Power is transmitted through a manually shifted, two speed, power glide automatic to a 4.89 ratio rear end.

CNG equal to 2 1/2 gal of liquid fuel is carried in a 1,600 psi cylinder in the rear of the car. Obviously, range is not a concern in a race that is only 1/4 mile.

Fuel pressure is reduced in three stages from 1,600 psi to about 3 oz at the mixer/regulator that feeds fuel and air to the engine.

Fast performer

Mezquita says the car is very competitive. It runs the quarter mile in 9.9 sec and reaches a speed of 142 mph. The car can go faster, but that time and speed are limits set by the Super-Gas class.

Mezquita won Super-Gas Eliminator at the NHRA U.S. Nationals in Indianapolis last September and was No. 1 qualifier at the inaugural Heartland Nationals in Topeka, Kan.

Drag racing probably isn't the best proving ground to demonstrate the emissions reducing potential of natural gas.

But it demonstrates that CNG can work under severe operating conditions.

All reformulated fuels, alternatives, and vehicles will have to be thoroughly tested before the public's confidence in them can be assured.

A new hometown

John L. Kennedy **June 25, 1990**

Beginning with this issue, you'll notice a change on this page.

After 80 years, Oil & Gas Journal has a new hometown.

It was moved to Tulsa from Beaumont, Tex., in 1910 when purchased by its present owners.

Now we're back in Texas.

While still headquartered in Beaumont, OGJ's predecessor, The Oil Investors Journal, opened an office in Tulsa because the Glenn Pool strike near there made Oklahoma the leading producing state.

For the past 80 years, OGJ has been based in Tulsa. We still have a large staff there, including editorial production — the group that pulls it all together each week. And printing and mailing are still done there.

But as of this month, the Journal's top editorial and publishing managements call Houston home.

Recent Shifts

For decades, the Journal has had a large staff in Houston, too. And editors and managers located in other cities spent enough time in Houston to qualify as "associate" residents.

Still, it was sometimes less than efficient to do what needed to be done with Gulf Coast companies and with firms around the world with U.S. ties.

Since the mid-1980s, the petroleum industry has become increasingly international — its activity, its technology, and its markets. There's no reason to think this trend will reverse.

At the same time, industry restructuring and consolidation have concentrated even more of the information and influence in the international petroleum industry in Houston.

It's not the only petroleum center that is growing in importance, of course. But we think Houston is now the best location for the Journal's editorial and publishing headquarters.

Closer contact

This move doesn't mean we'll ignore any region. And our editorial mission — to provide news, technology, statistics, and analysis of the worldwide petroleum industry — is unchanged.

An important reason for moving is to be able to carry out this mission more effectively. By serving Journal readers, we also serve Journal advertising customers. And by moving to Houston, we can all be in closer contact with a larger number of our customers.

Closer contact with more readers and customers. More efficient use of time and money.

Those are the demands made of companies in the international petroleum industry of the 1990s.

The Journal must meet these demands, too. We think having our headquarters in Houston will help.

We may move again in another 80 years. In the meantime, we're happy to be in Houston.

Siegfried the Wonder Bug

Bob Williams **July 16, 1990**

Siegfried the Wonder Bug is getting worried.

After 17 years and a quarter-million miles, my 1973 VW no longer exhibits its usual happy disdain for the events of the world.

(Why, you ask, would anyone drive the same vehicle — notably one described as looking like a pregnant rollerskate — so long?

Dismissing the obvious arguments of economics, sentiment, and packratitis, one is left with the rationale Detroit so fears: A car is just a box on four wheels that goes from Point A to Point B cheaply and does it again the next day.)

Anyway, two energy crises did not faze Siegfried, nor has lead phase-down. He still happily gulps leaded regular, and when that disappears he'll survive nicely on some version of Super Hi-Test Eco-Gas U.S. refiners are cooking up.

Why Siegfried's worried

But Siegfried is quaking in his pistons now. A slick, sharp-looking brochure landed on my desk this week from Nebraska.

It boosts the state as a nifty place for ethanol developers to invest — including a 20 cents/gal producer payment the legislature just tacked onto the state's ethanol incentive structure — because "the environment is the issue of the decade," and ethanol is what "America needs to clean up its air."

The brochure quotes President Bush's endorsement of corn ethanol and ETBE as good for energy independence, national security, trade deficit, farmers, taxpayers (?), and the environment.

Sorry, but Siegfried sipped gasohol twice accidentally and gagged both times, buckling, stalling, and conking like the drunken horse in the movie Cat Ballou until the next fillup of vintage leaded.

If you want to try a massively subsidized ethanol experiment, says Siegfried, check out Brazil.

Brazil's salvation?

An $18 billion program to mass produce ethanol from sugar cane was to be Brazil's salvation after the energy crises, slashing oil imports, saving plantations, and boosting the economy. Until 1989, nine of every 10 new cars in Brazil ran on it, and gasohol helped cut carbon monoxide pollution in the big cities.

But now, shops converting gasohol guzzlers back to gasoline are doing booming business in Sao Paulo. There are shortages in nine states after government-owned Petrobras sold its ethanol stocks and plantation owners shifted to exporting sugar. Distillers refuse to plant cane because the government has cut ethanol prices to $32/bbl from 1979's price of about $60/bbl. The gap between subsidies to distillers and pump prices is worsening the dire fiscal woes of Petrobras. Petrobas is hiking prices for ethanol, slashing its minimum blending requirement, and, word is, may phase it out soon.

The upshot? Brazilian ethanol producers are scouring the world for ethanol supplies as many Brazilian motorists abandon plans for local vacations for fear they won't be able to drive across some states for lack of fuel. Ethanol producers are even importing methanol to stretch supplies.

Environmentalists unhappy

That has the environmentalists (already unhappy over expanses of rain forest chewed up for sugar cane plantations) up in arms, claiming methanol is a dangerous, toxic substance.

As if to underscore their claims, a methanol tank truck recently spilled its cargo near a Brazilian river town, tainting drinking water and resulting in fish kills and sick humans and livestock.

Gasohol for the Wonder Bug? Methanol?

Siegfied will walk first.

Stealth drillers

Bob Williams **August 6, 1990**

Oil operations should be not seen and not heard.

Sometimes that seems to be the crux of the problem the industry faces in California, where public opposition stems not from any genuine environmental fears but from the Nimby (Not in my back yard) syndrome. In other words, do what you want, just keep those tacky pumpjacks/platforms/refineries, whatever, out of sight. Don't spoil the viewshed, dude.

This is not cynical whimsy but a semilearned observation resulting from study of quasiscientific data gathered in almost a decade of covering California's oil industry.

It begs a larger question: Is there something in the psyche of Californians that inherently resents intrusion of an industrial presence upon the landscape — an intrusion, in oil's case, that has little choice of location?

DOG history

That's always been my suspicion, and now there is more evidence to support that view, in a new book commemorating the 75th anniversary (Aug. 9) of the California Division of Oil & Gas and available through the agency. It's entitled "Drilling Through Time: 75 Years with California's Division of Oil & Gas." Written by the dean of West Coast oil writers, Bakersfield Californian columnist Bill Rintoul, the book is a well-written, entertaining collection of history and legend and yore in a state where the size of the oil resource is matched only by the size of the opposition to its exploitation.

Rintoul devotes a chapter to industry's struggle over the decades to maintain a low profile while operating in the Golden State since the clutter of the Signal Hill boom. In addition to well-known projects such as the "exclusive condo developments" (what the camouflaged drilling islands off Long

Beach usually are mistaken for), he cites:

• A Union Oil official's offer of a new hat to any complaining clubwomen who could spot an oil well in the company's new camouflaged, muffled, drilling island in Sansinena field in 1949. There were no takers.

• The deeper pool strike on the 20th Century Fox studio lot in Beverly Hills oil field that had Universal Consolidated Oil erecting its own "sets," drillsites landscaped to resemble grounds of an estate.

• Signal's efforts in the 1950s to develop Cheviot oil field, in the shadow of John Wayne's home and 100 yd from an exclusive golf course clubhouse, by hiring Hollywood soundstage staff, planting concealing trees, and painting derricks to match the grass and sky.

• Occidental's 1966 erection of $1 million derrick disguised as a 10 story office building to develop an oil strike in the Beverly Hills trend.

• Mobil's "lighthouse" drilling structure at Venice Beach, a 1966 discovery.

Get the idea? Apparently what California oil companies need to drill offshore is a "stealth" platform that is the antithesis of the U.S. bomber: invisible to the naked eye but registering strongly on radar to avoid being a navigation hazard.

Out of sight, out of mind. . .

Why bother

John L. Kennedy **August 27, 1990**

How much oil does the U.S. import from Iraq and Kuwait? How does it get to the refinery? Why do gasoline prices go up immediately and fall slowly?

Where is Kuwait?

That's just a sampling of the questions asked of the Journal since Aug. 2 by reporters and other persons trying to get a handle on the likely consequences of events around the Persian Gulf.

There were others, too:

Why not convert most automobiles to electricity? What happened to the U.S. government's big push to produce alternate fuels — from the sun, from coal, from biomass? Why not just open the valves wider on U.S. wells?

Tough questions, all. Legitimate questions.

This time was different

That's why the Journal has had a long-standing policy of responding to inquiries from radio, television, and newspaper reporters.

If we are asked — and if we are able — we provide statistics, an explanation of industry operations for background, or an on-the-air interview.

Normally, this means an occasional brief TV or radio comment on the implication for gasoline buyers of a jump or a slump in the price of WTI.

But the past 3 weeks have not been normal times.

First, we had our own information gathering to do. We had to do some last minute rewriting for the Aug. 6 issue, then get started on a comprehensive story for the next week.

That's no problem. That's what we do.

Record number of requests

What made these few critical days different was the large number of requests for information, the vast majority from the general media.

Our editors tried to field them all, from short appearances on the evening TV news, to 20 min on evening drive-time call-in radio shows, to detailed statistics and maps of the Middle East.

Some editors got a dozen calls to provide information or do on-air interviews the day after the invasion.

Our Oil & Gas Journal Energy Database received at least 70 requests for detailed industry data the first 2 days of the crisis, plus 30 more requests for a map quickly produced to help answer reporters' questions.

Late the following week, this pace slowed some. But it didn't end, by any means.

It all takes time and effort. And communication isn't always totally effective.

But we think this effort is a good investment. Explaining to oil and gas consumers the challenges facing industry has never been more important. In this latest crisis, several of the inquiries I fielded indicated both considerable understanding of the U.S. energy situation and a willingness to learn more.

Of course, we'd love to spread these requests over a more orderly schedule. But the last 3 weeks have shown again that isn't the way the real world works.

Middle East datelines

Jim West **September 24, 1990**

At least 150% of today's news, it seems, flows from events in the Middle East.

Since Iraq's Aug. 2 blitz of neighboring Kuwait, news wires coming into Oil & Gas Journal's offices have hummed with stories of all imaginable sort but with a common theme: the crisis.

From news sources throughout the U.S. there is no end of information on the Middle East, what the crisis means, and how it affects the business world and other segments of the public.

Here's a sample of what we receive, with datelines:

DENVER — Micromedex Inc. reported searing heat and the threat of chemical warfare are not the only dangers troops face in Saudi Arabia. A number of poisonous snakes and scorpions creep about the desert. Among the inhabitants no one would enjoy meeting are saw scaled vipers, puff adders, and Arabian cobras.

COLLEGE STATION, Tex. — Stored in a library basement at Texas A&M University are 500,000 Nazi war documents that could help provide an alternative to Middle East crude. Scholars say the papers outline "secrets" of how Germany made oil from coal as the size of its war machine and Allied bombing depleted conventional oil supplies. The catch is that less than 10% of the documents are translated.

NEW YORK — Consumers, wary of things like rising gasoline prices, will spend more of their money on activities at home, said Thomson Consumer Electronics Sales & Marketing-Americas. Thomson figures Americans spent $9 billion last year on movie rentals in an effort to convert their living rooms into private theaters.

WASHINGTON — The Renewable Fuels Association said the 900 million gal of fuel ethanol produced in the U.S. this year will cut crude oil imports by more than 40 million bbl. "Furthermore, it said, "if we were to increase the use of ethanol threefold we could eliminate all of our yearly oil import needs for gasoline from Iraq!"

HERNDON, Va. — The Persian Gulf crisis and the Clean Air Act are moving the U.S. closer to a "postgasoline" era, the Automotive Information Council reported. The era will feature vehicles equipped to run on one or more fuels such as ethanol, methanol, compressed natural gas, and electricity. Electrically powered vehicles have been around for decades, but progress is slow. General Motors has developed a battery powered car capable of 100 mph, but it has a couple of drawbacks: a maximum range of only 120 miles and an 870 lb battery that would have to be replaced every few years.

Elsewhere, one analyst warns that anyone who hopes the Middle East crisis will lead to a renewed, highest level push for a U.S. National Energy Strategy aimed at reducing the rate of growth of oil imports is bound to be disappointed. Here's why:

WASHINGTON — A big eruption in the Middle East could rapidly send crude oil prices to $50/bbl and set off a national effort to reduce U.S. vulnerability to imports, said Melvin A. Conant, president of Conant & Associates Ltd. But with every week resulting in more firepower arrayed against Saddam Hussein, an Iraqi strike against Saudi Arabia or Israel, for example, would almost certainly be defeated. Oil prices might then reach $50/bbl for a day or 2, then slide to $25/bbl or less. With that lower price and no other threat to oil supply, how could an NES that requires things like energy conservation and higher taxes get American support?

Conant concludes: "If it takes a war and prolonged shortfall in supply to provoke the U.S. into a major, long term effort to reduce its energy vulnerabilities, the current crisis in the Middle East is unlikely to provide the reason."

No news is not good news

L.R. Aalund **October 1, 1990**

Telling journalists there is information they can't have is the incentive that cranks up their energy and ingenuity.

Nevertheless, many, including this editor, bent their lances on the Communist bloc. The toughest nut of all was East Germany.

As refining editor in the early 1980s, I took an interest in the German Democratic Republic (GDR), which Oct. 3 goes out of existence as Germany officially unifies.

For years, the GDR was supposed to have had a satisfied populace that ran one of the most efficient Communist economies. The country's gross national product, after the Communist bosses sealed the border and built the Berlin wall, ranked fairly high in the world. GDR athletes did churn out a lot of Olympic gold.

But our main interests there were technological, such as refining catalyst developments and advances in turning lignite into transportation fuels and chemicals. The price of crude oil at that time was zooming upward, and the world was seriously considering all forms of alternative energy.

Letters to GDR technology groups or questionnaires like the one used for our Worldwide Catalyst Report were never answered. Personal contact with them at the huge West German Achema chemical exhibitions got plenty of promises but no articles or information.

Even going there didn't work. A day with an arrogant GDR trade official in East Berlin was spent largely in a tour conducted by an atheist of some magnificent churches and a sport and swimming complex. The perplexing thing was that they did not open up even though they badly wanted to sell technology for dollars and deutschemarks.

Now the nut has been cracked, the veil of secrecy lifted. This editor has just returned from a fascinating trip there, and the information flowed. Newspapers are springing up all over East Germany. Thanks for this go indirectly to hundreds of thousands of East Germans who protested for several nights in the streets of Leipzig not quite a year ago and broke the Communist hold.

There apparently were three reasons for East German secrecy, aggressively enforced by a rather nasty state security organization. The reasons: ignorance, incompetence, and embarrassment.

One assumes that in an authoritarian government the people at the top know all. They just aren't sharing it.

The truth was that this government, and others like it, are as ignorant as everybody else. They don't have the foggiest idea what is going on. Those at the bottom of the pyramid cook up phony statistics and production reports to meet quotas on paper, while those at the top concoct and publicize numbers and developments for propaganda.

Their embarrassment came from knowing that their operations in general were obsolete and some of the sloppiest in the world, fouling the air and water of the country and causing sickness.

Those are the reasons they kept their mouth shut and did not have visitors. Unfortunately, coverups by some, here or there, won't fade away with the GDR.

Not a dull business

L.R. Aalund **December 24, 1990**

Writing and editing articles about petroleum refining must be a pretty dull occupation in the mind of the general public.

Refining rarely makes the evening news. And as something to do, writing about it must rank near studying chemistry in high school. Yuck! would be the average American teenager's reaction to that.

But this editor is here to give witness that this is certainly not the case and that it is an exciting job. But to believe that and that this editor is not really a nerd, the American public will have to understand that gasoline doesn't come from oil companies or service stations, which are often on the evening news. It comes from petroleum refineries. Once that is clear the man on the freeway may also see some real drama out there in the worldwide refining industry.

For one, take the fact that Iraq, which has only about 300,000 b/d of refining capacity (about half of that in seven dinky plants), gobbled up about 800,000 b/d of capacity in four (three of which are sophisticated) refineries when it invaded Kuwait. Whether the Iraqis can run these refineries or whether there are enough Kuwaitis and expatriates around to do it for them is uncertain. But that is academic because most of this capacity was dedicated for product exports, which aren't going out now. Because of this, somewhere — perhaps Singapore, Netherlands, or Norway — a refinery is pushing through just a bit more crude oil than it would normally handle at this time.

OGJ's international issue next week will contain the latest capacity information on all the refineries around the Persian Gulf, all of which may soon be on the screens of bombardiers. So unless a formula for peace is found, many more refineries around the world, including some in the U.S., may have to work harder next year.

Another bit of drama that is more welcome is the article on two big refineries in eastern Germany in this issue. One of those is nearly a shrine for industrial organic chemists. Publication of an article like this only 18 months ago would have put everyone involved in it in a police state jail. Now it is possible, and what is most heartening is that managers and engineers at these refineries have so quickly learned to talk openly and freely.

Finally, U.S. motorists will in the next few years be participating in one of the world's great air pollution control experiments. As the news report on initial results of an expensive auto/oil industry study of reformulated gasoline reveals, this is indeed an experiment.

Reformulated gasoline, which is lower in volatility and loaded with oxygenated molecules, is supposed to reduce smog. But there is a question whether it will help when burned in old cars.

The U.S. has plenty of old cars. Unocal Corp. reports there are 400,000 cars built before 1971 operating in Los Angeles alone. The fumes from the worst of the polluters were rich enough to run a Ford Taurus at cruising speed, Unocal says.

Unocal, as part of an innovative approach to air pollution control, bought 8,400 at $700 apiece and scrapped them. The payment could help the owners buy post-1975 autos with cleaner burning engines.

Unocal Chairman Richard J. Stegemeier reported last fall that scrapping those cars removed nearly 11 million lb of pollution from Los Angeles air this year. He also said it would cost $150 million, or 25 times more money, to reduce unwanted air emissions by the same amount from the company's Los Angeles refinery.

George Bush slept here

L.R. Aalund **March 18, 1991**

Meetings at the Oil & Gas Journal are infrequent, informal, and relatively short.

We don't have any committees, although we do sometimes have teams working on editorial projects. A lot of internal business here takes place in hallways, on the telephone, or by fax.

However, every few years we shut things down for a couple of days and gather for a full fledged editorial conference. Everyone involved in writing, editing, and presenting copy for OGJ attends. Last month we had our first conference in Houston, where our editorial headquarters are now located.

The meeting took place in a relatively small, nearby, woodsy hotel and conference center, The Houstonian, where our out of towners were housed. President Bush also lives there when he is in town. His permanent suite is, we presume, more spacious than the singles we reserved. This is his legal residence, and he must return to Houston to vote.

He is rarely here, although he apparently thinks of Houston often and steers some good business to the city, like last summer's economic summit and next year's Republican party nominating convention. Houston's Democratic mayor is elated.

We don't know whether Bush gets the corporate rate at The Houstonian, but Kim Jordan, OGJ editorial assistant who arranged the facilities for our conference, says the price for us was right. The place has good meeting rooms and food.

On the serious side, we laid plans at the conference to work more closely and efficiently together and discussed the changing oil world and its influence on our future editorial pages. New ideas were floated, some of which were vigorously shot down. One such comes vividly to mind.

The conference also was a good opportunity for editors who knew each other only as voices on the telephone to meet face to face. Two, in fact, hardly knew anyone, voice or otherwise. Keith Rappold and Anne Rhodes, respectively our new drilling editor and refining editor, had joined the staff just a couple of weeks prior to the conference.

The icing on the cake was three speakers we invited to keep us looking outward. They certainly did:

Patrick J. Early, president of Amoco Production Co., shared Amoco's views on natural gas, a resource in which his company holds one of the world's largest private positions.

George Reese, energy specialist with Ernst & Young, gave us a closeup view on doing business in the Soviet Union, where he now heads his firm's Moscow office.

And John Dosher, senior vice-president of The Pace Consultants Inc. and in this editor's opinion one of the most interesting and stimulating energy analysts around, spotlighted some probable and unusual developments.

Here are a couple in line with the focus of this refining issue. Dosher said just as soon as the Middle East war ends — the ground war was a little more than 2 weeks away at the time — a worldwide boom in refinery construction will start that will tax the capacity of the engineering and hardware suppliers. There possibly will be a shortage of engineers.

In the U.S., reformulated gasoline will pose enormous strains, so much so that some refiners may go out of the gasoline business or consolidate with others. In short, some of the refiners in this year's survey may not be here in 5 years.

Eclipse of U.S. gas prices

A.D. Koen **July 15, 1991**

Last week's solar eclipse grabbed headlines around the world.

But the almost total eclipse of gas prices on June 24, when Nymex gas futures fell to $1.118/Mcf, was far more alarming — for many U.S. producers.

With Nymex gas futures at a record low, many operators took a hard look at their bottom lines. However, accountants warn it is unsafe and too scary to look directly at bottom lines during a gas price eclipse.

Instead, producers are advised to look as far into the future as possible, when supply and demand are expected to come into balance and gas prices are expected to rebound.

An operator who doesn't want to wait for the future can:

- Cut a small hole in one side of a cardboard box.
- Cover the hole with black paper or aluminum foil.
- Poke a pinhole in the paper/foil.
- Tape a balance sheet inside the box opposite the pinhole.
- View the eclipse by holding the box over his head, standing with his back to gas prices, and allowing market indicators to pass over one shoulder and fall onto the balance sheet.

The box method has the advantage of inverting the bottom line and obscuring the lasting harm low prices are likely to inflict on U.S. gas reserves and production.

The popular explanation for gas price eclipses holds that excess supplies drive prices downward and encourage producers to accept whatever price is required to move their gas.

However, scientists studying low gas prices say the phenomenon is too complicated to explain with simple supply and demand. They say gas price eclipses occur when too many people position themselves between end users and the wellhead, setting up a sequence of transactions that conceals finding and production costs.

Conversely, from umbra — the darkest part of the eclipse's shadow — gas producers can see no part of their markets. U.S. producers viewed domestic markets from this perspective during the recent gas price eclipse.

From penumbra, observers can see part of their markets. Canadian producers selling gas into U.S. markets viewed the eclipse from penumbra and thus received higher prices for their production.

For the most part, solar eclipses are brief and unavoidable. Evidence suggests gas price eclipses are quite the opposite.

Investors viewing a price eclipse on gas markets generally become uninterested in underwriting exploration for more gas reserves or drilling of more gas wells. Consequently, gas producers are likely to reduce drilling budgets, lay off employees, freeze or cut salaries, or curtail production of higher cost reserves.

So far, the partial eclipse lingering on U.S. markets indicates that such responses are having little effect on gas prices.

More ominous are indicators that surplus gas suppliers are likely to flood U.S. markets for several heating seasons to come.

If so, even the box method of viewing the bottom line isn't likely to sustain many U.S. gas producers until healthy profits return to U.S. markets.

A Dutch treat

L. R. Aalund **July 22, 1991**

Why in the world were about 100 refiners and catalyst experts and 16 mighty plow horses gathered during the lunch hour on a rain and wind lashed Dutch beach early last month?

They (the humans) had forsaken the elegance and warmth of the nearby Scheveningen Kurhaus Hotel to witness a rerun of a myth-shattering, dramatic, scientific experiment that had first been demonstrated publicly in 1654 at Regensburg, Germany.

There Otto von Guericke, one of those private renaissance researchers of many talents, demonstrated he could create a vacuum and that air pressure was a powerful force. He chose his day well, when the mighty of the empire had assembled in a diet or "Reichstag." He proved to the highnesses there that two teams, each of two, four, and finally eight horses, hitched to two half spheres held together by a vacuum and pulling against each other could not separate the ball. But afterward, when a child opened a small stopcock the half spheres fell apart. The philosophical and physical ramifications of this event, however, went much further.

What Guericke really wanted from this "show" was, among other things, to publicly strengthen his theory that outer space was devoid of air. He did that and shot holes in some ancient philosophy and astronomy.

But back in the Netherlands about 3 centuries later, Dr. Ditmar Schneider, a lecturer in technical history at the University of Magdeburg, Germany, has replicas of the original half spheres and will travel. As part of the package, Schneider will lecture at varying length (in German) on Guericke's work and the man, a student of astrology and mathematics, an engineer, builder, inventor, and for 30 years mayor of Magdeburg.

Akzo Chemie engaged Schneider to enlighten and entertain the some 300 persons, including this editor, who attended its catalyst conference at Scheveningen early in June. Only a portion braved the beach.

Akzo had to furnish the horses, which, according to Cees Smit, one of the company's catalyst experts, were animals still used for plowing in the clay soil of the Dutch province of Zeeland.

Herman Lovink, Akzo conference chairman, jokingly (perhaps) said he was convinced that modern day horses could outpull their ancestors, and it would take only a total of eight steeds to separate the sphere. Schneider disagreed.

Eight failed on the first try. Even with 16 for the later try it was touch and go, with one team threatening for a moment to pull the other to Amsterdam before the sphere popped, filling the air with a strange odor of some singed substance.

This success was fortunate because there would have been no child around to open the stopcock, just a group of drenched adults.

Deja vu all over again

Bob Williams **August 5, 1991**

To quote the eminent American philosopher: "It's deja vu all over again."

If one's taste runs more to a more literary vein than Yogi Berra, there's George Santayana: "Those who forget the lessons of history are condemned to repeat them."

However it's put, that's a common feeling in the oil and gas industry these days.

After riding the crude oil price rollercoaster of the 1980s, the market's queasiness had hardly subsided when the 1990s were ushered in with yet another wild price ride, courtesy of Saddam Hussein.

North American natural gas producers, after several years of proclaiming imminent demise of the gas bubble and resurrection of $5/Mcf prices, now are just about ready to cast pig's entrails to divine January's heating degree days while fighting that "Haven't we been here before?" sensation as spot prices slide.

Next oil surprise?

If you revel in deja vu, take a look at what's happening in the Soviet oil industry and read The Prize, Daniel Yergin's epic, captivating tale of oil's role as a catalyst in modern history. Yergin, president of Cambridge Energy Research Associates, says the next "oil surprise" may come not from the Middle East but from the Soviet Union. Talking to Swiss bankers in Zurich earlier this year, he said, "We are once again at one of those moments in time — as in the 1870s, 1920s, and 1950s — when developments in the Russian and Soviet oil industry can have dramatic international implications.

"The effects of postwar capacity constraints on Middle East oil supplies could be felt much more strongly if Soviet production and exports continue on their downward course and if at the same time, economic recovery in the U.S. and elsewhere leads to stepped up oil demand."

Yergin contends risks of oil market problems in 1992-93 are growing, citing in particular rising U.S. oil imports and plunging Soviet oil exports.

Noting the Soviet oil industry "is in big trouble that is getting bigger," Yergin cites as key issues for that industry labor unrest and investment decisions in Moscow.

What many industry observers have seen as a litmus test for Moscow's oil investment decisions is the long awaited conclusion of an agreement between Soviet officials and Chevron Corp. on developing supergiant Tengiz field. That deal has stalled as this magazine goes to press.

Soviet oil as catalyst

Soviet oil as a catalyst for markets has ample precedent.

As Yergin writes in The Prize, living and working conditions in the Caucasus, home of the Russian oil industry, were deplorable early in the century. Conservatives decried the rapid growth of an oil industry fed by foreign investment. Baku became the hotbed of revolution, giving rise to such Bolshevik leaders as Stalin, who masterminded oil workers' strikes. Strikes led to open rebellion and ethnic conflict as Tatars attacked oil industry facilities, torching oil wells.

An earthquake followed, and in short order, two-thirds of Russia's oil wells had been destroyed and exports collapsed, Yergin wrote. Violent upheaval had again interrupted vital oil exports.

Here we go again, folks: Keep the safety bar down, hands inside the car, no standing. And stock up on Dramamine.

Staggered deadlines sometimes staggered

Bob Tippee **September 2, 1991**

People often ask how Oil & Gas Journal gets all those up-to-the-minute news stories, detailed technology articles, and in-depth special reports researched, written, edited, illustrated, printed, bound, mailed, and delivered, week in and week out.

The secret: staggered deadlines, rigidly enforced.

OGJ editors and artists operate under the stimulation — laypeople call it pressure — of weekly deadlines. These are points in time by which staff members must complete their work and after which the unfinished pursue other employment opportunities.

In a perfect world, all deadlines for each issue would occur all at once, say the day before publication. But physical practicalities must be accommodated: page-making, printing, binding, and so forth. They take time. They cannot occur all at once. Hence, staggered deadlines.

Deadlines for those parts of the magazine for which immediacy is paramount — the news section, newsletter, and editorial, for example — occur 4-5 days prior to publication date. Deadlines for other parts — technology, exploration, the cover — come 2-3 weeks earlier.

Usually, everything hums steadily along. Sometimes, the system yields to circumstance.

At the start of the Houston business day Aug. 23, a Friday, International Editor Roger Vielvoye called from London with troubling news: The concrete gravity base under construction for the Norwegian North Sea's Sleipner A platform had sunk that morning near Stavanger. Fortunately, there were no injuries.

With the Aug. 26 issue already on the press, OGJ's first chance to report the incident would be Sept. 2. That gave Roger until Aug. 28 to file a news story and until Aug. 29 to change and update his report — a luxurious amount of time in the OGJ news section.

As you may have noticed, however, the Sept. 2 issue of OGJ also contains the Offshore Northern Europe Report, which Roger had written and filed more than a month before the Sleipner mishap.

When Roger called about the accident, his special report, having met the demands of much earlier editorial deadlines, was deep in the maw of the page-making and printing process. The Sleipner project, of course, figured prominently in this story on Norwegian activity.

Furthermore, the cover photo, by then ready to print, was a night shot of the ill-fated Sleipner A platform's concrete gravity base, still quite upright and afloat.

So much for staggered deadlines. Now, everything had to happen at once.

Red Beans, Rice, and Tabouli

On Friday, Presentation Editor Bob Lair and Assistant Technology Editor Bob Lawson shifted the production process into reverse, retrieving the dated Norway story and cover.

Editors in Houston sent a new cover photograph to Tulsa for quick processing, and the Sleipner photo passed to the news section to illustrate the Sleipner accident story.

On Monday, Aug. 26, Roger sent Houston a new Norway story for the Offshore Northern Europe Report for editing and copyfitting to replace the old story, line for line. Modern electronics then conveyed the story back to Tulsa, where it eventually caught up with the special section and the rest of this issue of OGJ.

Learning Soviet culture

Robin Buckner Price **September 23, 1991**

The line of companies seeking business in the U.S.S.R. keeps growing.

Most are well aware of the difficulties ahead. Among the more often thought of hurdles are lengthy contract negotiations, lack of advanced technology, and lack of a monetary exchange policy.

But long before contracts are negotiated, equipment considered, or any money changes hands, one must meet with another human being, a Soviet, and get the ball rolling.

At this basic level are major cultural differences between the Soviets and their foreign counterparts.

The Soviet business world is full of long-standing traditions and practices.

Dealing with Soviets

To help meet that challenge, international accounting/consulting firm KPMG Peat Marwick and Sondra Snowdon, international protocol specialist and author, offer these tips for doing business with Soviets:

• Refer to associates as Soviets, not Russians. Russian is the most common language, but not all citizens are Russian.

• Send all letters of inquiry and requests by facsimile. Mail service to the U.S.S.R. is unreliable. Many Soviets have fax machines but it may take a couple of days to get a fax through. Soviets need encouragement to respond, so be persistent.

• When traveling to the U.S.S.R. take the names, phone numbers, and addresses of Soviet contacts. There are no Yellow Pages, and trying to find someone later can be difficult.

• Many Soviets don't have business cards because of the country's paper shortage. Carry many business cards and offer them to everyone attending a meeting.

• Being creative with currency can make a difference in a business deal. Hard currency is in short supply. One company is selling Soviet artwork outside the U.S.S.R. and using the currency to buy tennis shoes to send back to the artist.

• Invitations to visit your company outside the U.S.S.R. will be appreciated, but the host delegation should arrange to pay all expenses, including pocket money. Hosts also should take out health and accident insurance for visiting Soviets.

• Modest business gifts are widely accepted but avoid giving knives or letter openers because they signify the severing of a relationship. Flowers are fine, but yellow flowers symbolize betrayal.

• Soviets dress plainly, and foreign businessmen should leave designer clothes behind.

• Entertaining is important to a successful business relationship with the Soviets. Choose a round table in a restaurant because corner seats are considered unlucky. Offer the maitre d' a pack of cigarettes to ensure good service.

• Be prepared for lengthy toasts to friendship throughout the meal. The Soviet host will use either vodka or champagne. The proper response is to raise the glass and say "Nazdorovye!" (nazdah-ROV-yeh), meaning, "To your good health!"

While following these tips may not get a well drilled quickly, if Billy Bob Judd slaps a dozen yellow roses and a set of steak knives on the square table at his dinner meeting with Vladimir Zhyankaslovasaksomova, chances are Billy Bob will go home empty-handed.

Some of the biggest opportunities for the petroleum industry are in places where the course is uncharted or vaguely defined at best. China, Yugoslavia, Albania, Romania, all offer great potential for people willing to take time to understand the people, then focus on business.

Meantime, Nazdorovye!

Do not fold, staple, mutilate

Anne Rhodes **October 14, 1991**

Among many challenges an OGJ technology editor faces is the seemingly simple but often perplexing task of obtaining a clear, undamaged photo of an author to run with his or her technology article. It is our policy to run such pictures along with a synopsis of the author's professional experience.

We deal with a great many authors' pictures from around the world. In 1990 we published articles by about 395 authors. While acquiring such photos — passport size or larger — is usually not difficult, getting a printable version is sometimes maddening.

Paper clip woes

Paper clips are often the culprit.

For example, a package recently arrived at one editor's desk containing six author synopses, or "bios," and photographs. The pictures were of good quality, and the bios well written and of proper length. But each photo was paper clipped to its corresponding bio, leaving the authors with paper clips wrapped at varying angles around their heads.

In addition, one of the pictures was so badly scratched by the paper clip that it was unusable. A second copy of the photograph was requested. When it arrived, it was (again) paper clipped. It was however, paper clipped face down the second time.

The problem in this case is that a paper clip leaves an imprint on a photograph, making the author — as Bob Lawson, assistant technology editor, puts it — "look as if he parts his hair with a paperclip."

Identifying authors

Another time, a set of photographs arrived with the authors' names clearly written on the backs of the photos — in indelible ink. Apparently, photos were stacked on top of each other. Had the pictures been printed as is, each author would have appeared with another's name imprinted across his face.

On the other hand, sometimes there is no identification at all for pictures — particularly perplexing when you have more than one author. A simple telephone call can usually remedy this situation. However, in one instance, this was not the case.

Recently, a pair of unlabeled author photos were received with an article.

Each man wore glasses and a muted plaid sports jacket, and neither had any distinguishing features, such as a moustache, scar, or facial tattoo.

This required an especially tactful approach, in which case words such as "older" or "better looking" would be inappropriate. Fortunately, their ties were uniquely striped, and the photos painlessly identified.

There have also been cases of the photo being taped to a page, which usually results in part of the photo being removed with the tape.

OGJ photo rules

These remarks are meant to be educational, not critical. To that end, here are the first published rules for submitting photos to an OGJ editor:

- Do not paper clip, staple, tape, nail, or otherwise affix the photo to any other item.
- Always identify the person in the photo.
- Do not write directly on the back of the photo — or on the front, for that matter. Writing on the back of a photo can leave a visible shadow on the front.
- Write the author's name on an adhesive label before affixing the label to the photo. Then affix the label to the back of the photo.

Because these photos are important elements of an OGJ technical article, we in turn will observe our rule of treating these pictures with care and not losing or mixing them up.

Rape oil and crystalline rock

L. R. Aalund **November 11, 1991**

Only 5 or 6 years ago relatively few people, other than locals, knew where the small town of Windischeschenbach, Germany, was.

This editor certainly didn't but now does and can even spell it right four out of five times. Most important, Windischeschenbach is now well known in world scientific and geological circles because of the drilling rig on the edge of this charming town.

This super rig is drilling toward a depth of 40,000 ft (about 12 km or 7 1/2 miles) as part of the "KTB" research project funded by the German government. The rig's capabilities, progress, and objectives are described in an article beginning on p. 49.

The drillsite is in a region of high hills, dark forests, trout streams, and pretty villages. But the project, countryside, and recent history make this a region of paradoxes.

The Czechoslovakian border is only a few miles away. That, just a little more than 2 years ago, was a section of the Iron Curtain, making this part of Upper Bavaria for about 45 years a sort of dead end, a backwater for western travelers. It was a no man's land for daring eastern travelers. This border with its isolated woods and valleys was the favorite crossing point in the novels of a number of English spy novelists, who made a mint from the Cold War.

Perhaps for similar reasons, the U.S. Army established the largest infantry/artillery training reservation in Europe at Grafenwoehr. You can tell when you pass it by the signs along the road: "Keep Out — U.S. Army." On a summer evening one can hear the distant thunder of cannons amid the church bells of Windischeschenbach as a "firing exercise" proceeds into the night.

Generations of American GIs remember their winter training exercises at Grafenwoehr with something less than fondness.

Although one is practically in the geographical center of Europe, the autobahns and fast highways all seem to be flowing north and south, not east. Driving in that direction is cutting against the grain. No one wanted to give the Russian tank corps the use of an eight lane autobahn to Nuremberg.

A visit to the KTB rig floor with Claus Chur, the KTB manager who has been involved with the project since 1986, gives one a view of this panorama.

In the summer, the surrounding fields are a brilliant yellow from rape plant blooms. An edible oil is made from rape seed and, apparently because of some European Community incentive, farmers all over the continent are infatuated with this plant.

Of course, this has sparked a European project to convert the oil to "Biodiesel."

Publicity includes a nice ecological loop showing the plants using carbon monoxide from cars to make oxygen and ultimately fuel.

A young farmer interviewed by a local paper reported he experienced no drawbacks in running his tractor on the fuel, but he said it made his farm smell like french fried potatoes.

Rebirth of a nation

A. D. Koen **December 2, 1991**

Fahed S. Al-Ajmi in early August 1990 flew out of Kuwait City's international airport only hours before Iraq launched its surprise attack on Kuwait.

When news of the invasion caught up with him, Al-Ajmi tried to return to Kuwait City, but the airport had been closed.

A procurement officer of Kuwait Oil Co., Al-Ajmi was destined to fight the Persian Gulf war from Houston. He was a key figure in the Al-Awda restoration project, described by project manager Bechtel Inc. as one of the construction industry's largest, fastest mobilizations of personnel and materials.

Since the fighting ended, Al-Ajmi has returned to Kuwait four times, the last in October when the end of the wild well control effort was in sight.

Environmental damage

Kuwait's deserts remain fouled by lakes of oil and its coastline still is smeared by oil slicks. But at least, Al-Ajmi noted, the Kuwaiti people at last have smokefree air to breathe. A great deal of work remains to be done to clean up history's worst oil pollution.

Al-Ajmi said cleaning up Kuwait's environment is one of two priorities on the government's agenda. The other is restoring oil output at least to levels before the Iraqi invasion, an effort begun before the last well was ceremoniously capped Nov. 6.

For Kuwaiti oil officials, controlling Kuwait's 749 blowouts was just the first step in restoring a national oil industry devastated by Iraqi sabotage.

"Killing the fires and controlling the wells were simple tasks compared with what we're going to do next," Al-Ajmi said. "It's not going to be easy."

He contends KOC's most pressing task is to survey damaged wells to decide which can be salvaged, many of which will have to be abandoned and redrilled. As KOC evaluates the situation, it also is studying ways to improve its production efficiency, Al-Ajmi said.

New spirit

Al-Ajmi gives much of the credit for progress so far to the help from people and companies around the world. Also, he said, the Kuwaiti people have found a new unity and purpose in the task of rebuilding their country and society.

That in sharp contrast with a previously widespread image of the Kuwaitis as a wealthy, privileged elite dependent on government to take care of all their concerns.

"At one time we sat and wanted things from the government given to us. But now it's the other way around. There's something I've seen in Kuwait that I've never seen before. People are eager to work, to do something for their country."

Iraq failed to wipe Kuwait off the face of the earth by annexing it. Failing that, Iraq's efforts to destroy Kuwait's oil industry also proved unsuccessful.

One of Saddam Hussein's bitter charges against Kuwait was that the method of that nation's birth did little to confer legitimacy to the emirate's claim of nationhood.

Ironically, the Iraqi invasion served to spark a rebirth of Kuwaiti national identity and purpose, according to Al-Ajmi.

"I've never seen such spirit in the Kuwaiti people. So many are from different backgrounds, but we shared the tragedy and it has brought us together in a way we've never been before."

Bean counting in the oil and gas industry

Bob Tippee **December 16, 1991**

An Oil & Gas Journal staffer not long ago fretted publicly that her teenaged son was failing mathematics at school.

She had urged and encouraged, cajoled and threatened — to no avail. (Teenagers are like that.)

So a staff curmudgeon offered this advice: Tell the lad that if he enters adulthood armed with no more than scoffing familiarity with mathematics he will look dumber to his associates than he would now if he showed up at the mall on Saturday afternoon with his mother.

Insensitive but true. (Curmudgeons are like that.) A person with a weak grip on mathematics has no way of understanding the swarm of numbers that — for good or bad — increasingly defines human activity.

To be sure, numbers aren't everything. In the modern, hypertechnical, superdigitized, and micromanaged world, they're probably on the order of 75% of everything. And the share is growing.

This certainly is so in the oil and gas industry. As the popular complaint goes, the bean counters have taken control. More precisely, everybody's a bean counter now.

For that reason, Oil & Gas Journal is expanding its back-of-the-book Statistics department — by 11.5% in normal, 52 issue years, for anyone keeping score.

The third issue of OGJ each month, beginning herewith, will contain three Statistics pages instead of the usual two. The extra page per month allows Economics Editor Robert J. Beck and Statistics Editor Joan Bonfield Biggs to introduce monthly data that until now had nowhere to go.

Appearing for the first time in this issue are data on gas processing margins from Wright Killen & Co. and OGJ estimates of crack spreads in the U.S. refining industry.

The additional page further provides regular space for data available only once a month. Before the expansion, monthly data had to preempt weekly numbers, so continuity of weekly data suffered.

To cite just one example, the "Gasoline prices" table until now gave way once a month to "Active rigs abroad." The gasoline price numbers were available on the week each month when they did not appear; space for them was not.

Now the monthly "Active rigs abroad" table will appear with all the other monthly tables on the extra Statistics page in the third issue of each month. All weekly data now will appear on the standard pair of pages without the monthly interruptions that occurred before.

Sound confusing? To tell the truth, it's simpler than it used to be. Just watch.

It's simpler than before because Bob and Joan know their mathematics and were able to decide what numbers are important and when they tend to show up, and could sit down with Presentation Editor Bob Lair and figure out ways to make it all fit into the magazine. Lair knows mathematics, too, but doesn't often admit it. (Presentation editors are like that.)

Even OGJ Editor John Kennedy — everybody's boss — employed the bit of mathematics he can remember when he decided that, yes, the costs of an 11.5% increase in the Statistics page budget were more than justified by the benefits to readers of more thorough statistical coverage.

If this is not a roaring case for mathematical education then no such case exists. Of course, none of it explains why teenagers, curmudgeons, or presentation editors are the way they are or why nobody likes bean counters.

As we said, numbers aren't everything.

Your prospect is great, but

G. Alan Petzet **March 2, 1992**

Consulting geologists trying to sell exploratory and development prospects to or evaluate prospects for oil and gas companies hear a variety of reasons for rejection.

Louis J. Mazzullo, a geological consultant in Midland, Tex., has compiled a list of his top 20 reasons for having a prospect turned down.

The West Texas Geological Society published Mazzullo's list in the February 1992 issue of its bulletin. Here's the list:

20. We just don't have any exploration (development) dollars left for this one.

19. We're elephant hunting, and this is just a field mouse.

18. There's not enough of an acreage position to make this worthwhile.

17. There's too much "goat pasture" acreage. You're trying to over-promote the prospect.

16. My boss once drilled a dry strat trap near Bangor, Me., and he's burned out on that type of prospect.

15. We are looking for a prospect where you can get updip of a 300,000 bbl well.

14. This prospect is downdip of a 300,000 bbl well. It must be drained.

13. We've just reorganized, and management hasn't decided what direction to take.

12. Not enough well control.

11. Too much well control. It should have been found by now.

10. There are no downdip shows of oil.

9. There are no updip shows of oil.

8. There are too many oil shows. It should have been found by now.

7. Sample shows in the key well are too good. The reservoir must be tight.
6. There is not enough seismic closure on that strat trap.
5. There is too much seismic closure. It should have been found by now.
4. My boss doesn't like — County.
3. We haven't run economics on this yet, but the terms are too tough.
2. Our plate is full.

No. 1 reason

And the No. 1 reason for turning down a prospect:

1. It doesn't fit our pistol.

Mazzullo said many times rejection is not based on any heady scientific evaluations. Many are polite ways of saying, "We wish we could consider your deal, but we just don't have enough (money) (manpower) (time) (experience in the area) [Pick one] to adequately evaluate it."

Mazzullo said he does not take offense when a deal is turned down unless his character is assassinated along with the rejection, but admits rejection can be a blow to the ego.

He noted he has used many of the same reasons when turning down deals screened for his clients. He said he did not present the list as criticism of industry policy but rather as a challenge to see how many reasons he has missed.

He said he had about 40 reasons by mid-February.

Faces behind the numbers

Bob Williams **March 9, 1992**

Tracking oil and gas production numbers is an important task for Oil & Gas Journal.

That's because they are the most obvious yardstick of the current health of the oil and gas industry, notably when they are well out of line with demand numbers. Variables affecting a nation's oil and gas production are too numerous to mention here, but now and then a new one pops up.

Russia's energy crisis

Russia, for now the No. 1 oil and gas producer, is grappling with crises in its energy sector. But Russia's natural gas industry has been a relative bright spot, with output stable in 1991.

However, says Soviet Business & Economic News (BizEkon), Moscow, Russian gas industry officials fret over their industry's future. Market dislocations from free market measures are a concern. The price of gas in Russia rose only fivefold last year vs. a tenfold jump in the cost of a gas processing plant and a thousandfold leap in the cost of pipe for gas pipelines, BizEkon quoted Moscow independent newspaper Nezavisimaya Gazeta as reporting. Russia soon will allow as much as 50% of gas output to be sold at free market prices.

However, the Moscow newspaper says, the big worry is the flight of personnel from the gas production regions. In the construction sector alone, the number of workers in western Siberia has plunged to 20,000 today from about 90,000 in 1987-88. One reason is the prospect of higher paying jobs in the newly emerging private sector in Central Russia. The other is the ethnic makeup of the Siberian gas industry workforce. Ukrainians account for 30-50% of Tyumen workers. With the breakup of the U.S.S.R., they no longer need rubles because Ukraine is moving to create its own currency. In addition, Russian law provides for an earlier pension age than does Ukraine law. In fact, says the Moscow newspaper, if Ukrainian workers in Tyumen take only Ukrainian citizenship, some areas in western Siberia won't have enough Russian citizens to elect local governments. Another worry for Ukrainian workers in western Siberia is Ukraine's courting of Iranian oil and gas. If Kiev is looking outside the commonwealth for other oil and gas supplies, will it be as interested in the problems of Ukrainian workers in Tyumen?

There's a similar dilemma with the big Tatar community in Tyumen. Further, Tyumen authorities have yet to hash out jurisdictional differences with the government of the Yamal-Sakha republic, and gas workers don't know whom to report to.

A human face

There's a human face behind the vagaries of oil and gas production statistics. Sometimes it's a face that buttresses the headlines: Yamani, Khomeini, Saddam, Bush, Gorbachev, Yeltsin.

Sometimes, it's a weatherbeaten, anonymous face reflecting the rigors of working a brutal job under miserable conditions for wages that buy less than they did yesterday paid by an authority that may not be there tomorrow.

Anonymous, yes. But for want of a horseshoe nail . . .

Life in the downsized lane

Jim West **April 13, 1992**

One of the words that appears most often in press releases flowing across Oil & Gas Journal's news desk these days is "downsizing."

Synonyms are "streamlining" and "rightsizing."

No matter what the action is called, it means lots of people are losing their jobs because of a recession that simply won't go away despite economists' assurances that it is, predictions that it will, or warning that it won't. Whatever.

Readers of all sorts of publications have a wide range of choices among analyses of what's happening in today's economy and why.

Employees caught in layoffs know exactly what's happening. Perhaps their most intense interest lies in why. "Why me?"

Cutting by headcount

Cutting the workforce by the number won't save U.S. industry, warns Work in America Institute Inc. (WAI), Washington.

Companies claim to be reducing their ranks in the name of strategy.

"But the connotation of broad, clear, long term objectives seldom fits reality," WAI said. "In most cases the so-called strategy is a desperate, short term effort to cut costs now and worry about the future later.

"Rather than a strategic withdrawal, downsizing is too often a retreat that leaves little or no basis for a counteroffensive."

WAI acknowledges that some cutbacks may be required to trim bloated corporate bureaucracies. But cutting by headcount fails as strategy because it devalues vital human resources, especially managerial and technical backbone.

There is an implicit assumption that downsizing selectively screens out less productive, lower achieving workers, WAI says.

"This certainly would be desirable, and it is possible. But as generally implemented, downsizing is at best nonselective. It is at worst selective in the wrong way."

Here's why:

Many companies achieve reductions in force largely through voluntary early retirement and severance packages. Those aren't controlled separations affecting only the lower end of the performance curve. They target the most experienced segment of the managerial and professional talent pool.

What's more, they often attract a company's best and brightest, leaving it lean for sure but stripped of much of its best talent.

"The result," WAI says, "is an organization that is more stressed for leadership and effective management and consequently less able to compete effectively."

No immunity

Regardless of size, no organization is immune to downsizing, says a study by Right Associates, Philadelphia management consultant.

Released last month, Right's study of 1,204 U.S. companies showed 73% or 909 companies of all sizes underwent personnel layoffs during the last 5 years. Among downsized companies, 26% had fewer than 500 employees, 14% had 500-999, 29% had 1000-5000, and 24% had more than 5,000 employees.

"Downsizing obviously has become a tool not just of corporate giants where massive layoffs make front page news but of thousands of average sized companies all over the U.S.," Right said.

Small comfort for employees who have been downsized and whose careers may be washed, as a friend calls it.

Roger Vielvoye 1941-1992

John L. Kennedy **May 4, 1992**

Sometimes he thought much faster than he typed. Sometimes he wrote more than we could fit into one story. Frequently, he caused editors in Tulsa or Houston to search the dictionary for words not common in southwestern U.S. oil field conversation.

Often, not a single letter in his handwritten note was completely legible.

But through the pages of Oil & Gas Journal, Roger Vielvoye helped our readers around the world understand the complexities of the international petroleum industry.

Last week, he died unexpectedly at his home near London. He was 50.

Roger had the unqualified respect of the engineers, managers, executives, and CEOs that formed his huge network of news sources. He had the same standing with his peers in publishing.

Journal career

As Journal international editor for 14 years, Roger traveled extensively in Europe, the Middle East, and Asia. From his base in London, he operated independently several thousand miles from the home office. Because of his knowledge and experience, he could choose the best stories to write and knew how to write them.

He generated ideas, discussed them with other Journal editors, then followed through.

On paper, it sounds like a dream job. And I have no doubt that Roger liked his work.

But even a dream job has demands. The job of international editor, which Roger molded, requires a deep and broad knowledge of oil and gas operations, activity, and technology in diverse geographical regions, an ability to glean from that knowledge what is most important to Journal readers each week, initiative, and hard work.

Roger had all this; most of all, he had initiative and hard work.

Respected professional

He was one of our most prolific writers. An example of his work leads off this issue's Offshore Report. But he did more than report and write. His interest was keen in any event or trend in petroleum or publishing that would affect industry's fortunes or the Journal's success.

He represented the Journal — and the industry — in the most professional way.

Being selected to receive the 1990 Offshore Northern Seas (ONS) Conference Press Prize was just one indication of the value of Roger's reporting.

". . . He has a broad and influential readership from Los Angeles to Leningrad, both in industry and government circles," said Frode Gjeitvik, ONS chairman at the time.

Roger's interests were not limited to petroleum operations. He could discuss a broad range of subjects from barbecue to cricket with intelligence and a finely honed sense of humor (or humour, as he would have spelled it).

Journal readers will miss his analysis of petroleum issues, events, and activity. His friends and coworkers will miss much more.

Cheers, Roger.

No pictures allowed

L. R. Aalund **May 18, 1992**

Journal editors and writers are chroniclers rather than historians.

In fact, if too much history showed up in our pages, we would soon be out of business.

Our business is, of course, to keep you and our other readers up to date.

But history can be useful and interesting.

These aspects prompted us to put together, with the help of four experts on fluid catalytic cracking (FCC), our special report tied to the 50th anniversary of this technology.

FCC arguably is the most important development in petroleum refining technology in the past 50 years. We hasten to point out, however, that the articles in this section are not totally focused on history.

How it was then

Nevertheless, Journal editors who worked on this report did mentally spend a lot of time in 1942, which was both a frightening and scintillating year. The Japanese had bombed Pearl Harbor in December the previous year, the world was learning where the Solomon Islands were, and World War II was building to a crescendo.

In the U.S., Enrico Fermi split the atom and Irving Berlin wrote "White Christmas."

The American South in 1942 was not the bustling "Sunbelt" it was to become but a region climbing out of the dirt-poor poverty of the Great Depression. But it did have oil and refineries.

Texas was the world's largest oil producer, and the great refineries on the Gulf Coast were turning out the fuel for distant land, sea, and air battles and feedstock for critical synthetic rubber. This, of course, drew German U boats to the Gulf of Mexico and caused the blackout of cities and refineries on the coast and well up the Mississippi.

John P. O'Donnell, former Journal editor who now lives in Houston, bylined an article from his office in Manhattan that appeared in our May 28, 1942, issue describing the importance of the new FCC technology in making gasoline, high octane alkylate, and butyl rubber.

Timely work, but few people, including Journal editors, knew then the first FCC had been started up only 3 days earlier at Baton Rouge. The first news of this — without site — would appear in OGJ's June 11, 1942, issue.

One obviously didn't walk in or near refineries with cameras in those days, which is apparently why we had difficulty finding construction pictures.

Volume 90

John L. Kennedy **May 25, 1992**

If you read the fine print, you'll notice that Oil & Gas Journal is now labeled Volume 90. In publishing jargon, that means we are in our 90th year.

It was 90 years ago this month that the Journal's predecessor, Oil Investors Journal, was first published. That was Volume 1, Number 1. When ownership changed 8 years later, the name became Oil & Gas Journal, and the frequency was changed from bimonthly to weekly. Since 1910, the name and the weekly frequency have not changed.

By fall of 1902, publisher Holland Reavis had 1,000 paid subscribers, and he quit distributing free copies. We've been a paid magazine ever since.

Most importantly, the editorial objectives set in those early years — accuracy and objectivity — still guide our writers and editors each week.

The next 90

On any anniversary, it's tempting to cite long-ago conditions and fabricate "lessons" based on the striking difference — or the striking similarity — between then and now. I won't do that. I'd rather speculate on the next 90 years.

Can world economic growth continue on a fairly steady course for 9 decades? If so, what will that mean for energy demand? Will efficiency gains offset demand increases resulting from economic expansion?

Will oil and gas still be significant energy sources in 2082?

Remember that for much of the past 90 years, forecasters soberly predicted the world would soon run out of oil. Yet reserves of conventional oil gained 50% in the 1980s alone. At current consumption rates, conventional oil reserves represent 45 years' supply. Adding other sources — including future discoveries — the total oil resource may be 8-10 times that amount.

The oil will be there; the question is whether or not we will use it.

Is there even a remote chance that gasoline will still be a significant fuel in 2082? It doesn't seem likely. Yet the internal combustion gasoline engine has been the leading source of power for individual transportation for 100 years. It has successfully fought off challenges from, for example, electric automobiles for almost as long.

What about power generation in the late 21st century? Will the world's ever-growing demand for electricity be met by nuclear power, or will natural gas finally be the utilities' fuel of choice?

Volume 180?

Will Oil & Gas Journal still be published in 2082?

Maybe. It's likely Holland Reavis, as he prepared the first issue of Oil Investors Journal, didn't expect it to last 90 years. And it wouldn't have without the loyal support of generations of readers and advertisers.

That's a lesson that has been passed along for 90 years at the Journal.

I've resisted the urge to concoct other lessons from history on this 90th anniversary of the Journal's start-up. But I'm not making any promises about future anniversaries. When we reach Volume 100 in 10 years, the temptation may be too great.

And for answers to today's questions, be sure to read Volume 180, Number 21.

Reformulated service stations

L. R. Aalund **September 14, 1992**

The end of the traditional U.S. driving season Sept. 7 and a recent column by Journal International Editor David Knott on food and drink retailing at British service stations has prompted some reflections on the U.S. scene by one harried commuter.

There is a lot of food and drink moving across the service station bays of the U.S., too. It is sad, in a way, to see the once grand surviving sisters of the oil world, symbolized by mighty tankers on the high seas and free road maps, now engaged in mulling their market share in beer and chips.

As a result, the hard sell is on at many of their emporia, which are selling just about everything from antacids to Zwieback. Oddly, the sale of gasoline seems almost secondary at some.

Handy quarts

A case in point is one of the majors in the Houston area, which was for months — and perhaps still is — into an aggressive lube oil sales campaign.

Typically at one of its outlets, one could rush up, late for an appointment, to the cashier in the glass cage to charge the gasoline and hear, even with a tie, white shirt, and just cleaned and pressed suit on, a cheerful: "Need any oil?"

"No thank you."

"We have premium 10W30 in handy quart sizes just behind you on the rack."

A touch of humor at this point may get things off high center.

"Well, I could use an oil change."

"Sir, this is a self-service station. We don't change oil. Anything else you need?"

"Just my credit card and ticket."

"Now don't rush me."

Just think. Some executive probably spent several years in an MBA program learning how to create that situation.

Snuff and such

Or there is the cashier at another major's station with fingers poised over the cash register, ready to ring up a cash gasoline sale.

"Need any thing else? Soft drinks?"

Again, try something a bit outlandish to get things moving.

"You have any chewing tobacco?"

"We've got what they call smokeless tobacco, or snuff, and loose Red Man and Beechnut in the paper pouches. But we don't sell any plug tobacco."

This information provides an out, an escape hatch. With hand outstretched for five bucks due in change, "Well, skip the tobacco. I kinda wanted the plug."

But wait . . . the cashier completes one more synapse honed at the company sales training sessions.

"Would you like to apply for a credit card?"

That's an idea. Perhaps one can even buy gasoline with it.

Continuing education offers some hope

Keith Rappold **September 21, 1992**

Why on earth would engineers in the U.S. return to school to study petroleum engineering?

Are these people pure academics, eternal optimists, or wild daredevils?

Many analyses of the upstream petroleum industry in the U.S. point to continued low activity. This bleak outlook indicates that pursuing an advanced petroleum engineering degree should be classed with other risky activities such as bungee jumping and skydiving.

Petroleum engineers have therefore entered MBA programs to expand their marketability, open doors in peripheral areas such as gas marketing, or prepare to enter management.

Field of choice

A few engineers, however, continue to pursue technology in their field of choice: petroleum. A genuine love of the industry is frequently coupled with the need to upgrade skills to remain employed.

Unfortunately for some, regaining employment has become the biggest challenge. An advanced degree and college placement services can help an engineer quickly reenter the oil industry.

In a University of Houston graduate petroleum engineering class, the reasons for pursuing an MS vary widely. Some engineers have received nontechnical or peripheral assignments following the most recent rounds of restructuring.

A drilling engineer no longer worries about logistics, penetration rates, and daily costs.

Last month, he became an environmental coordinator, and he spends a great deal of his time drudging through complex environmental regulations written by nontechnical lawyers, many of whom may not even recognize the outdoors much less the oil field.

Why does he study petroleum engineering? The classes stimulate his mind and keep him abreast of new technology, for perhaps one day he will return to the technical position he covets.

In contrast, a mid-level manager has been thrust back into a "grunt level" engineering position, and he admits his engineering and computer skills can use some honing. The need to stay abreast of technology becomes more apparent as he reflects on the high skill level of some of the recent college graduates fortunate enough to join his major oil company.

The few BS graduates, fully computer literate, know the industry is weak. Many have taken extra courses to gain a competitive edge.

Resting on laurels

A highly paid 10 or 20 year engineer cannot rely on past achievements. He must exhibit at least the same level of competence as newer engineers, as well as demonstrate practical knowledge from years of experience, to justify his position and salary.

Some larger companies have reduced training programs, leaving engineers to fend for themselves as technology advances.

Continuing education makes engineers more adaptable and offers flexibility for other assignments. In addition to technical courses, some part time engineering students have enrolled in foreign language and geography courses.

They realize all too well that staying in the oil industry may not mean staying in the U.S.

Chevron's offshore bargain shopper

Robin Buckner Price **January 18, 1993**

Who would have thought bargain shopping for soda pop could save a company enough money to — perhaps — retain those two geotechs or a few of those administrative assistants who were lopped off in the last restructuring?

Jerry Couey, that's who.

Couey is a production operator for Chevron Corp. on a South Timbalier platform off Louisiana.

He has designed a system he calls Chevron Abandoned Purchase Paradigms (CAPP) that saves Chevron an average 35% on all items purchased.

Couey's CAPP

Couey's CAPP system sidesteps middlemen by buying items needed on the platform direct — direct from the Wal-Mart discount store in Leeville, La., just 12 miles from Chevron's shore base.

Couey's system is described in the November/December 1992 Chevron Focus.

A list of items needed, like paper products, hand tools, and soda pop is faxed to Wal-Mart before a platform crew change. Wal-Mart checks off what is available on the list and faxes the modified order back to the rig.

The new crew picks up the items on its way to the rig.

Couey discovered by shopping Wal-Mart he could save Chevron 63¢ on enamel spray paint, $14.96 on wire cutters, and $35.06 on a 3/8 in. ratchet drive.

Chevron used to pay $60.34 for a combination wrench set but now gets the same item for $9.24, a big 85% savings.

"When you consider how many of these items we go through, it definitely adds up," Couey told Focus.

Couey figures his system yields savings of $68,000/year on soda pop alone.

Couey says the key to his system is there are no long term, no-cut contracts.

"If we give all our business to a single company, they can eventually jack up prices on us while they snuff out their competition," Couey said. He wants to see a system company-wide in which two bids are required for every order.

And with Wal-Mart, buying is done order by order, so if an item is not available or is cheaper somewhere else, Chevron can buy it somewhere else.

"We're trying to abandon some of the long-standing purchasing practices that aren't exactly cost effective," Couey told Focus.

"We've been watching the pennies, and now the dollars are starting to roll in."

Bargain shoppers

Bargain hunting for supplies and equipment. What a novel idea.

With oil company earnings in the cellar, it seems about time to start penny pinching.

And one would think a company doesn't necessarily need to start cutting employees before looking at what it spends on soda pop.

Sure, cutting 500 employees with an average salary of, say, $35,000/year will do a lot more for the bottom line than shopping Wal-Mart, but if three, or two, or even one employee gets to keep their jobs because a company started shopping prudently, isn't it worth it?

One could imagine that perhaps Couey has for years watched his company pay double, sometimes triple what he pays for the same item when he goes to Wal-Mart to get something for use at the house.

And one day perhaps he just asked, "Why?"

Here's OGJ's position

John L. Kennedy **March 8, 1993**

Several times a year I speak to petroleum industry groups, usually about petroleum supply and demand and industry trends.

Of the questions posed by the audience at these presentations, those that involve oil and gas activity, technology, or markets are the easiest to answer.

Some of these speeches — and answers to the questions that follow — include a dose of speculation. In mid-February, at the Oil Industry Taxation seminar during Institute of Petroleum's IP Week in London, I outlined what I thought would be President Clinton's energy tax plan.

After some presentations, I am asked to explain the Journal's editorial position on a current issue. A recent example: "Tell me why on earth the Oil & Gas Journal doesn't support a U.S. oil import fee!"

In whose best interest?

Following a speech late last month in Houston at the Petroleum Equipment Suppliers Association's 1993 Human Resources Forum, I was asked, "Does OGJ take positions that are in the best interest of the industry or in the best interest of the country?"

Good question.

The quick answer is that we don't think there is a conflict between the two. Petroleum is the lifeblood of a healthy economy, so a healthy oil and gas industry is in the best interest of any nation. Oil & Gas Journal has its own interest, too. And we know from experience that successful oil and gas companies and successful supply/service companies serve that interest best.

The complete answer is more complex. Oil & Gas Journal covers worldwide petroleum industry operations from seismic analysis to gasoline blending. Not all our readers agree on what makes industry healthy.

Opportunities for a clash of interests abound — between oil producers and gas producers; between domestic firms and multinationals; between producers and refiners; between operators and supply/service/contracting firms; between gas producers and gas pipeliners; between integrated companies and independent producers or refiners; between large and small.

I could go on. And within each group, there often is a clash between short term advantage and long term health.

A framework for all

Those who serve only one of these segments sometimes assume it is easy to see what is in that group's best interest. Sometimes it is easy; often it is not. Some who serve only one group believe the best interests of the various groups are widely divergent.

We don't think so.

We think the interests of all industry segments can be accommodated in a framework that permits the long term value of petroleum to be optimized and allows market forces to determine energy prices and choices.

That's our position. We strive to make sure the action we call for on any issue fits within this framework.

Ethanol ad nauseum

Anne K. Rhodes **March 22, 1993**

Of all the hardships facing refiners today, transportation fuels requirements are receiving the most attention.

Oxygenated and reformulated gasolines, legislated as part of the U.S. 1990 Clean Air Act amendments, are aimed at reducing atmospheric carbon monoxide and ozone levels in major U.S. metropolitan areas.

The oxygenated gasoline program began last December in 39 U.S. cities with carbon monoxide problems, as determined by the Environmental Protection Agency.

MTBE

The oxygenate of choice among refiners and gasoline marketers is methyl tertiary butyl ether. Methyl tertiary butyl ether is manufactured by reacting methanol with isobutylene, and business is booming. Oxygenate capacity more than doubled during 1992.

But if current events are any indication, the future for MTBE producers is, at best, uncertain.

Complaints by consumers in Alaska have halted sales of gasoline containing MTBE in Fairbanks. Alaska Gov. Walter Hickel banned the new fuel in December after 120 motorists complained of headaches and nausea, allegedly caused by gasoline containing MTBE.

The Centers for Disease Control in Atlanta reported that several Fairbanks residents had measurable levels of MTBE in their bloodstreams. Residents of Denver and Las Vegas, however, who have been using gasoline containing MTBE for several years, have not reported such ailments.

Mary Smith, director of EPA's field operations and support division, questions the validity of the claims. "I fill my car with [gasoline containing] MTBE," says Smith, "and I haven't felt any ill effects."

EPA is also investigating several complaints from consumers in Colorado, Montana, and New Jersey and hopes to decide by November whether it should reconsider the use of MTBE in gasoline.

Thanks, guys, but it's a little late for refiners.

Ethanol

Smith believes Montana's recently expressed preference for using ethanol to oxygenate gasoline is based on its status as an agricultural state. This seems to be corroborated by Mike Bryan, a spokesperson for the National Corn Growers Association, who said, "There very clearly could be a problem with MTBE."

Can you say, "special interest?"

And in the latest and perhaps most demoralizing move from a refiner's perspective, the White House is contemplating exempting politically powerful ethanol from energy tax plans.

With heavy hitters like Archer-Daniels-Midland Co. and the Fertilizer Institute continuing to hammer on farm-state lawmakers, refiners may be left in the dust with unneeded capacity to produce an unwanted chemical.

Anyone else starting to feel nauseous?

Wisdom at OTC's 25th birthday

Bob Tippee **May 10, 1993**

Landmark birthdays make you older and wiser, or just older.

The Offshore Technology Conference turned 25 last week. Matthew Simmons, who runs Simmons & Co. International, brought wisdom to the party.

In a press conference, Simmons compared the offshore oil and gas industry of 1968 with the modern version. Then he described what the industry might be like when OTC turns 50.

In 1968, he said, 70 drilling contractors owned and operated all 233 rigs in the worldwide offshore drilling fleet. The top 17 contractors owned two-thirds of the rigs.

Ten of those 17 contractors are gone or are owned by someone else. The new top 17 own only half of today's fleet, which is three times bigger than it was in 1968.

Jack-ups' share of the fleet has grown to 60% from 38% in this period. Semis have booted tenders out of second place, their fleet share doubling to 24% while tenders' share dropped to 5% from 23%.

Water depth

Naturally, the most impressive changes concern the water depth in which the offshore industry can conduct its business.

In 1968, Simmons said, the water depth record for successful exploratory drilling was about 700 ft. Now it's more than 7,500 ft.

Production occurred in water no deeper than 250 ft in 1968. Today, there's a platform in 1,800 ft of water in the Gulf of Mexico, and subsea wells produce in more than 3,000 ft of water off Brazil.

Simmons went on to show how the government of one country, the U.S., has profited from the offshore oil and gas business.

By 1968, the industry had paid $1.9 billion to the federal government for Outer Continental Shelf leases covering 5.5 million acres. Since then, it has paid $58 billion in lease bonuses for another 58 million acres.

The government makes money from royalties and taxes, too, of course. During 1953-67, total government revenues from OCS oil and gas activities, excluding associated income taxes, totaled $2.8 billion. The figure for the next 25 years is nearly $100 billion.

"When you realize how much dependable and cheap energy the offshore industry added to our country, let alone another $100 billion in revenue from just lease bonuses and royalties, one would logically conclude that the offshore industry would be one of the U.S. industrial heroes, like the industry is in Norway and the U.K.," Simmons said. "Sadly, this never happened."

Ignoring the record

Simmons blamed an environmental movement that ignores the record: Only 126,500 bbl of oil seeped or spilled from producing wells and 980 bbl from exploratory work during 1971-90. Simmons calculated that the government has gained $740,000/bbl of oil spilled.

"If any other industry sector could begin to match this record, this type of economics would go a long way toward cleaning up most of the worst pollution throughout the country," he declared.

Maybe the U.S. will wise up in OTC's next 25 years.

Aramco's respect for oil

L. R. Aalund **June 21, 1993**

It seems as if most documentaries — mainly educational TV presentations — or exhibits focus on contributions wind, solar, hydropower, coal, or biomass will make to mankind's progress into the future.

Any attention devoted to oil is almost historical in tone, with dinosaurs often favored for the background motif. Horizontal drilling or catalysts have yet to be invented in those presentations. It's almost as if all those jets up there winging moms, dads, and kids from Liverpool, Osaka, or Duesseldorf to Disney world in Orlando are burning pink lemonade.

Oil is, of course, not "cool" in many circles. It's even a bit disreputable, given its alleged greenhouse association. But the fact is oil is the only source of energy that can keep pace with the world's burgeoning population in the foreseeable future.

Oil does get the respect it deserves at one place. That is at the Aramco exhibit in Dhahran, Saudi Arabia that I had the pleasure to visit while on assignment gathering material for the Middle East report.

World's finest

The Aramco exhibit is housed in a bright, airy, distinctively designed building of glass, polished marble, and stainless steel. Its area is considerably larger than an American football field, with 6,600 sq m or 74,000 sq ft. The exhibits cover about one third of the area, or 26,000 sq ft.

Surrounding a core area that illustrates the rich Arabic/Islamic technical heritage are seven display clusters on the modern petroleum industry, beginning with an introduction to petroleum. This is followed by displays on the origins of petroleum, searching for petroleum, drilling, producing, and reservoir management, oil refining, gas fractionation, and the transportation network. A final cluster is designated the "Aramco Story."

Many of the displays are interactive with the viewer and are actuated and controlled by him or her pushing, cranking, turning, or touching. Directions, comments, questions, and answers are in Arabic or English via video or voice.

Kids are fascinated with all this, but it is not "kid stuff." The exhibits, for example, make the immensity of the Ghawar field oil reservoir, the world's largest, comprehensible. They also explain crude oil distillation, describe the origin of oil, or show an ancient calculator.

Even the meaning of API gravity is clarified for the layman. Each of four closed cylinders filled with oil has a disc at the bottom. When a child or adult turns a crank all the cylinders are quickly turned end on end and the discs, now at the top of the cylinders, begin to fall to the bottom. Their speed down is proportional to the density of the oil. The one in the Arab light crude, of course, wins the race to the bottom while the one in Arab heavy is still at midcylinder.

A road show

It would be a major enterprise to put this exhibit on the road for some of the world's oil consumers to see.

Perhaps a better idea would be to send the U.S. president, vice-president, and entire Congress over there for a weekend.

It would be tax dollars well spent.

Left side vs. right side

A. D. Koen **July 19, 1993**

Conflicting attitudes in the U.S. about energy, the environment, and the economy are taken for granted by many.

But not by this editor's mentor, Max Flow.

As Max sees it, the paradoxes run deeper than simple differences of opinion among oil and gas producers, environmentalists, and politicians to the cradle of thought itself: the human brain. According to Max, American opinion splits over how to produce enough energy to expand the U.S. economy while protecting the environment because the brain's left and right hemispheres naturally have conflicting thoughts.

High school biology

As anyone who studied high school biology knows, the left side of the human brain thinks logically, organizing time and events sequentially, defining and cataloging objects, following rules, and setting goals and acting. The right side of the brain is more imaginative and intuitive. It synthesizes data: trees into a forest or oil fields into trends, provinces, or structures.

While the left side of the brain moves methodically, step by step, the right side can leap toward understanding in sudden, unexpected flashes of insight.

But can Flow's Dictum about the brain's cognitive strategies explain paradoxical American views on energy, the environment, and the economy? Consider these examples:

- The left side knows natural resources in the U.S. are scarce and social needs pressing, but the right side wants to spend whatever it takes to protect the environment.

• The left side of the brain knows alcohol fuels, biomass, and solar, wind, and tidal power together account for less than 0.3% of U.S. energy and without government subsidies would be virtually uncompetitive in domestic markets. Yet the right side of the brain wants to expand the U.S. economy while reducing dependence on fossil fuels.

• The left side knows the U.S. since 1989 has been importing more than 2 million b/d of oil from the Middle East, up from about 472,000 b/d in 1985. But despite recurrent turmoil in the region, the right side is willing to continue increasing U.S. reliance on oil imports, while discouraging domestic E&P.

• The left side wants to protect U.S. coastal environments from oil spills, while the right side wants to continue blocking offshore drilling and production, thus assuring the U.S. will need more tanker borne oil.

Get the picture?

The left side knows environmental regulations should rely more on physical science and sound economics. The right side is content with regulations based on political science and sound bites.

Fusing thoughts

Clues to Flow's left brain/right brain theory have been there all along. It just took someone of towering intellect — like Max — to synthesize the logical.

With the left side of this editor's brain satisfied about the root of U.S. ambivalence toward energy, the environment, and the economy, the right side leapt quickly to a possible solution: Could the U.S. energy-environment-economy conflict be resolved by fusing the brain's two hemispheres? The idea has individual if not social merit.

So let's end the debate. Who wants to be first?

The power and the glory

Anne K. Rhodes **October 11, 1993**

The South Texas wind was blowing hot dust around, even though it was late September.

The 90 deg. F. heat felt even hotter in a flame-retardant, distress-orange lab coat.

One thought kept returning during this refinery tour, arranged for a small group of editors and reporters: The potential energy around here is almost palpable.

Someone said, "We're running a little behind schedule. I guess we'll have to skip going to the top of the cat cracker."

Was this some kind of cruel joke? Had they really intended to take us to the top of that processing unit that resembled a small skyscraper? It was, after all, the dominant feature of the Three Rivers, Tex., skyline.

"Lloyd sure will be disappointed," she continued.

Lloyd is Lloyd Grubbs, superintendent of the fluid catalytic cracking (FCC) unit at Diamond Shamrock Inc.'s Three Rivers refiner. With the recent addition of a hydrocracker, not to mention the FCC unit, this small, relatively new refinery is well equipped to convert crude oil.

Grubbs has put in more than a few years in this industry. He has operated a cat cracker — although not this one — for about 42 years. The process itself is only 50 years old.

Fourth floor, please

As it turned out, they weren't kidding. They really had planned to take us to the top of the cat cracker. This was one opportunity that was not going to slip by.

After only a little coaxing, Lloyd enthusiastically herded us into a freight-type elevator and took us up about 200 ft. There we stood near the top of the reactor.

Steep, narrow stairs led up farther, to the top of the catalyst regenerator. Now the energy really was palpable.

The rapidly circulating, hot catalyst roared through the heavy steel equipment. The structure vibrated.

The cyclones, which separate the catalyst from the oil, were at arm's length. Lloyd beamed with pride as he led the group around the back of the regenerator. Waves of heat emanated from the vessel.

Enlightenment

It is one thing to write about petroleum refining and compile catalyst surveys. It is quite another to feel the dynamic power of the single most important refining process.

Suddenly it became clear why people like Lloyd are proud to have transformed oil into gasoline for nearly half a century.

Wanted: a few good quotes

Bob Williams **November 1, 1993**

The petroleum industry generally has done an inadequate job of getting out its side of the story on environmental controversies in which it gets embroiled.

So those of us at OGJ who track such things are especially cheered when industry does find its voice, notably when environmental extremism lurches past problematic into the surreal.

Chevron recently said the Environmental Protection Agency's stop work order on a reformulated gasoline project at its El Segundo, Calif., refinery over a permit bobble was "bizarre" and "makes no sense to us." Unocal Corp., finding itself in similar straits, called EPA's similar action aimed at it "an abuse of the permit process," and said EPA "tried to run roughshod over Chevron in a similar manner."

Earlier, EPA hit Chevron Chemical with a $17 million fine over a paperwork violation Chevron said involved harmless substances. Chevron said it was "outraged" and the action "illustrates a federal regulatory process run amok."

Wanted: a few good quotes

Those are good quotes, in plain English, the typical daily newspaper reader or broadcast sampler can understand. No PR puffery or "No comment" or long-winded technical jargon sprinkled among corporate hemming and hawing that's been blessed by counsel.

What value does that have? A survey by American Opinion Research (AOR), Princeton, N.J., for the Foundation for American Communications shows that despite surging demand for environmental reporting, U.S. journalists are not doing a good job of covering the environment. Who says? Just the 512 journalists surveyed. Fewer than one in 10 said the press is perceived as a very good source of unbiased environmental information. And 72% said reporters lack training and background to cover stories on technical environmental issues. Previous AOR surveys showed 81% of the public relies on the news media for environmental information. Almost half the journalists said their organizations have expanded environmental coverage the past 2 years, but only 21% of editors and 30% of television news directors said general assignment reporters — many of whom get the green beat — are well equipped to cover environmental issues.

Reporters admit bias

The study also found:

- Many journalists have an anti-business bias, said 36%.

• Almost seven in 10 surveyed said they have difficulty finding unbiased environmental experts. Almost six in 10 say it is difficult to find experts who speak "plain English."

• Reporters are much more likely to get story ideas from the government or environmental activists than from business sources.

AOR said, "In every market we've studied in the past year, environmental news has ranked in the top 25% of topics of interest but in the bottom 25% of satisfaction with the job the press is doing."

Can we say "Opportunity"? "Initiative"?

I sure hope that isn't the sound of hemming and hawing I hear.

The other Santa Barbara soap opera

Bob Williams **December 13, 1993**

There recently was on U.S. network television a short lived soap opera called Santa Barbara.

For the persevering folks at Exxon Co. U.S.A., there has been a long running soap opera of the same name that just recently closed its most important chapter.

For Exxon, the Santa Barbara soap opera was a $ 2.6 billion project whose progress was measured in almost geological time and whose plot had more twists and turns than a daytime drama.

Santa Barbara saga

Exxon discovered reserves estimated at 500 million bbl of oil and 1 tcf of gas underlying the Santa Ynez Unit in the Santa Barbara Channel in 1969, just months after the Union Oil blowout and spill whelped the modern environmental movement. It wasn't until earlier this month that Exxon was able to fully develop and produce those reserves.

The battle for permits to fully develop Santa Ynez reserves raged for more than 25 years from hostile local turf to the pinnacles of U.S. government, with plenty of court visits. It included human drama as well, such as the Exxon executive who at a turning point in the fight told the Santa Barbara County Board of Supervisors to "stick it (the permit) in your ear" and the marriage of an Exxon employee and a county employee who fell in love over the bargaining table.

In true soap opera fashion, Exxon began as the "villain" of the piece but ended with a "model project," in the words of local, state, and federal officials. Despite all the heated rhetoric, phalanxes of lawyers, and sometimes nasty political battles, Exxon eventually will be producing 80,000-90,000 b/d, and pretty much everyone that counts has kissed and made up.

A model project

In fact, Cynthia Quarterman, deputy director of Minerals Management Service and a Clinton appointee, said, "I'd like to use this project as a model for future conflict resolution," adding that MMS Director Tom Fry has asked her to spearhead an alternative conflict resolution program that will deal with U.S. controversies ranging from offshore oil leasing and development to royalty issues.

At the risk of sounding cynical, I wonder what kind of a model a 25 year permitting battle represents. Still, it's the results that count, and that can serve as a model in itself.

Even more encouraging is the trend this represents, as can be seen in other initiatives involving Offshore California's vast untapped hydrocarbon resource.

So the plot thickens: Will a new paradigm of industry-government conflict resolution emerge from California, previously the seedbed of such conflict? Will industry make more progress in developing oil and gas off California? Will there ever be, heaven forbid, another lease sale off California?

Stay tuned.

A letter from Nigeria

Patrick Crow **July 4, 1994**

I recently was faxed another letter from Nigeria.

This one, in awkward English, was from the recently widowed "Mrs. Mary White." She explained that a friend in the U.S. consulate had given her my name and vouched for my "honesty, trustworthiness, and maturity of handling the business."

She claimed to need my help desperately. She had documents proving her right to $37 million in her late husband's bank account, but some of his Nigerian counterparts are fighting over the money.

She asked if she could transfer the money to my bank account, then travel to the U.S. to collect it.

Here's the catch

Mrs. White wrote, "If you assist me urgently as anticipated, 25% of the total money would be given to you. All I need from you are your bank particulars—bank name, account number, bank address, your private fax, and phone numbers too, for privacy."

Her signature was oddly masculine.

Of course, if I had complied, I would have quickly lost what little money I have in my bank account.

I receive a couple of similar Nigerian letters every year because my name and fax number are printed in Oil & Gas Journal.

The U.S. Commerce Department gets hundreds of complaints a year about the Nigerian flim-flam industry. Some of the scams are quite complex and succeed occasionally.

I recently read a newspaper story about a Virginia businessman who was sent one of these letters and decided to out-con the con men. He traveled to Nigeria, discovered he was up against some tough thugs, and was lucky to escape. More recently, a South Dakota church was informed a Nigerian benefactor had willed it $3 million, but it had to pay a $90,000 inheritance tax first. Unfortunately, it did.

A few years ago, when oil was a more precious commodity than today, a frequent swindle was to offer people outside the oil industry the chance to buy a bargain-priced tankerload of Nigerian crude, said to be en route to the U.S. without a buyer. OGJ's Washington office received several telephone calls from excited people asking about recent prices for Bonny light. After I warned them about Nigerian ripoffs, their moods changed to embarrassment.

Amused, insulted

Mrs. White's letter triggered a mood change for me, too.

At first I was amused at the absurdity of the offer. Then I became annoyed that she had insulted my intelligence and my journalistic savvy with such a clumsy ruse.

Unlike other recipients, I can exact a little revenge for Mrs. White's letter. I can write about her and the Nigerian letter rackets, hopefully warning any of our readers who never have received such a letter.

If you receive one, don't let yourself enjoy the crude psychology and the inventive English too much. There's nothing funny about these letters. The writers are thieves who prey on naive people, and it's a shame they continue to flourish.

Medics and bodyguards

L.R. Aalund **August 29, 1994**

Surprisingly, Helsinki, with its modern infrastructure and amenities, may become a base for explorationists headed for Siberia.

This is one reason Juhani Missonen, managing director of Euro-Flite Ltd., Helsinki, was at the Offshore Technology Conference (OTC) in Houston last spring. We set the wheels in motion then to meet members of his staff and associates.

In addition to charter business flights, Euro-Flite jets medical teams into Russia and other former Soviet Union (FSU) countries to evacuate seriously sick or injured personnel working for oil companies. They are flown to medical facilities in Helsinki.

Euro-Flite has signed a 3 year contract to evacuate personnel associated with the Pechora group (Amoco, Exxon, Norsk Hydro, and Texaco) working north of the Arctic Circle.

The flight to this region takes over 3 hr by jet from Helsinki. Russian hospitals cannot deliver advanced medical care. Some travelers to the FSU carry their own hypodermics and antibiotics.

Ground risk

Providing this air ambulance service poses risks bigger than flying in arctic conditions.

Arto Johansson, flight operations manager for Euro-Flite, and Matti Siimes, captain of its Falcon 20 jet, give fascinating sketches of flying in Russia.

Siimes is an articulate veteran pilot with a demeanor that would instill confidence in any passenger. He says there are hundreds of airports in the FSI, but few people know where they are because they formerly were military secrets. Also, although flight controllers are good, they rarely speak English. A Russian navigator can be a necessity.

Another major problem is that airport managers demand cash for landing fees and fuel. As a result, pilots must carry $ 5,000-10,000. Desperate individuals and organized, violent criminals know this. In fact, any visitor from the West is considered rich and fair game, whether in Archangel, St. Petersburg, or Moscow. Hotels often tell their guests not to speak English in public.

These dangers created the need for bodyguards for flights or for business passengers who want them. The guards will even spend the night in the hall at your hotel room door. They come from Helsinki but are licensed to carry weapons in the FSU. These professionals probably don't need them, being masters of martial arts.

No safe place

Siimes took this editor to a warehouse-like building near the lively Helsinki waterfront where Euro-Flite's security service has its bright offices and a spacious gymnasium for training and private instructions. He introduced Slava, a key staff member, who told us about security in the FSU.

Question to Slava: "Helsinki seems to be a safe city, with many people walking the streets at night. If you rated it a '10,' what would you rate Moscow?"

His reply: "You are a journalist. I am a security man. I don't compare cities in that way. For me, there is no safe place."

Max's new offshore project

A. D. Koen **October 3, 1994**

I found my mentor, Max Flow, humped over his desk, working furiously in the dingy, cluttered broom closet he uses as an office.

When casting about for ideas or seeking new perspectives, I often visit Max. His vast store of knowledge about petroleum and mankind can provide just the sort of insight an editor needs to break that next blockbuster oil or gas story.

A chain of truth

Max motioned for me to sit down amid the brooms, mops, and pails. Turning, he asked, "Grasshopper, how many oil and gas platforms are in the Gulf of Mexico?"

I was sure he knew the answer, but the master often poses such queries before stringing together a chain of universal truths.

"At least 3,800," I responded confidently.

"And how many tons of shellfish are harvested each year from platform jackets in the gulf?" Max persisted.

This time he had me.

"And how many offshore platforms in the gulf have been converted to seafood restaurants after their wells were plugged and abandoned?" he continued.

Max read the puzzlement in my eyes and pressed on.

"Haven't you heard about the reception fine restaurateurs on the West Coast are giving to mussels and other shellfish harvested from the jackets of Shell Western E&P Inc. platforms off California?" he asked incredulously.

I had to admit I hadn't.

Shell, Max explained, has trimmed platform cleaning costs by 75% with a mariculture project in which a contractor periodically scrapes shellfish from Emmy, Ellen, Elly, and Eureka platforms off Huntington Beach, then sells the harvest to discriminating restaurants. It turns out the blue Mediterranean mussels among the hauls are especially coveted for their size, flavor, and cleanliness.

Shell's 4E shellfish are cleaner because the water around the platforms is cleaner than near shore, where water has been sullied by runoff from California's sprawling urbanization. Also, because the shellfish don't grow on the ocean floor, they can't collect sand from the bottom, another plus for diners.

"My plan is to start a similar maricultural operation in the Gulf of Mexico—sort of a Rigs To Restaurants program — but go one step further and eliminate the middleman," Max confided. He cast conspiratorial glances around his tiny cubicle. "Why take the harvest to restaurants on shore? Platforms have galleys, don't they?"

Four Star restaurants

"I'll buy a few, convert them to Four Star eateries, and serve the freshest shellfish on the Gulf Coast!" Max beamed triumphantly.

"But mighty one," I protested, "it's never been done. What will state health officials say, and the Food & Drug Administration, and your liability insurer?"

"Mere technicalities," Max growled. Then he dashed out the door on the way to his new project, scattering brooms, mops, and pails in his wake.

A brutal reason for restructuring

Bob Tippee **November 7, 1994**

If you work for an oil and gas company of any size, prepare for more change. Restructuring has not run its course. It may never end.

This quickly becomes clear in the API Report in this issue. Life in an oil and gas company half a decade hence won't be what it used to be or what it is now.

What may not yet be clear to everyone is the reason for all this change: The oil industry has been destroying economic value.

Grim numbers

Analysts at McKinsey & Co. Inc., New York, estimate that between 1980 and 1993 a sample of 103 worldwide oil and gas companies vanquished nearly $300 billion in shareholder wealth, mostly in upstream operations.

The analysts measure this as the difference between actual and expected market performance. Expected performance is what share price gains with dividends reinvested would have totaled, adjusted for risk, if they matched overall market performance in the companies' respective countries.

Extrapolating these findings to all nongovernmental oil and gas companies yields a total worldwide loss of wealth during the period of $450 billion.

The totals exclude Royal Dutch/Shell Group and Exxon Corp., which created "substantial value for their shareholders" during the last decade, the analysts say.

In a McKinsey Quarterly article, Charles Conn and David White blame the industry's grim decade on maturation of producing fields, declining discovery size, environmental regulations, and increasing government takes from production.

Costs rose while prices for crude oil, adjusted for inflation, fell. And things haven't improved much.

Job cuts, asset restructurings, reductions in exploration costs, moves to new areas of exploration, and technical innovations have been impressive. They just haven't been enough.

What the changes have accomplished for the average oil and gas company is survival. Companies, say Conn and White, remain "high-cost players in a low-cost game." Most of them still destroy wealth.

What to do? Change games, say the McKinsey analysts.

Past, future success

Successful companies of the past decade combined "a deep understanding of industry dynamics with their own special capabilities to create distinctive, difficult-to-imitate strategic responses to the key questions facing the industry," Conn and White say.

Successful companies of the future will follow a similar course: finding opportunity in "discontinuities" such as local market imbalances and breakthrough technologies; building "capability platforms" from strategically bundled people, management processes, and assets; expanding capabilities incrementally; and motivating it all with "genuine passion, commitment, and long-term vision from senior managers."

Other prescriptions exist, of course. It's the diagnosis that matters most. An industry that cannot create wealth cannot endure.

That's why things can never be the way they used to be.

The forecast season

Bob Williams **January 2, 1995**

'Tis the season to predict oil and gas prices.

Maybe it's all the eggnog at Christmas or the champagne at New Year's.

Whatever the reason, the yearend holidays seem to bring a bagfull of forecasts, predictions, prognostications, guesstimates, and entrails to be divined.

You don't need a crystal ball or a dart board, just a good grounding in history. OGJ last month received from Gary R. Jackson, a partner in the Midland, Tex., law firm of Jackson & Williamson, a copy of a talk Midland consulting economist Jim Williams gave at a Texas Society of CPAs conference. The law firm had engaged Williams to prepare an oil price economic study.

Jackson writes of Williams' study, "I found it fascinating to observe the relative consistency of the values and the ratio of the median and average oil prices, in 1993 dollars, for the different time periods presented."

Williams concluded, "Over the long term, oil prices have averaged near $18/bbl in 1993 dollars. But this number is somewhat deceiving, since prices have been below $15/bbl over 50% of the time. The only times since World War II that prices have exceeded $20/bbl were during a war in the Middle East. With this history in mind, oil producers should probably expect prices near $15/bbl and should be able to operate below that price for sustained periods of time."

Jackson adds, "It is depressing to realize that in 1993, using 1981 dollars, the price of oil was $9.07/bbl, and the price of natural gas was $1.28/Mcf."

Which is why this column was not written for last week's issue. We wanted to be sure no one would be reading it the morning after New Year's Eve.

Season's greetings.

Beethoven's last symphony

L.R. Aalund **January 16, 1995**

Now that management consultants have directed the "right sizing" and "downsizing" of just about everything in the oil industry, they are looking for new prospects.

Sports and the fine arts could be targets.

For example, replacing 11 man football with six man football in the U.S. National Football League would save millions of dollars in salaries. Six man football is played where schools don't have enough boys to field an 11 man squad, like in parts of West Texas.

On the other hand, NFL teams with their special teams need about 50 highly paid players. Six man would cut the team size nearly in half and still provide a serious but wild and woolly game that pro fans would love. What's good for Amoco, Mobil, West Texas, and the rest is bound to be good for the NFL.

A symphony?

But is it possible to downsize a symphony orchestra? We have seen an anonymous report, a spoof (we hope), from Europe about a prestigious consulting company that wants to do just that to a famous German philharmonic. The company is as well known in management as McDonald's is in hamburgers.

Here is what its head MBA advised the conductor after hearing one symphony:

- The oboes are silent too much. This is a waste of personnel, and they should play more.
- The 12 violins all play the same thing. This is unnecessary duplication of work. The size of this group should be drastically cut. If higher volume is needed, an amplifier could be brought in.
- There is too much repetition of certain passages. It does not make sense for the horns to repeat a passage that has already been worked on by the violins. The score should be reworked with this in mind.
- All 32nd notes should be changed to 16th notes. This would permit the orchestra to use music students and cut costs further.

If these and other measures are instituted, the consultant estimates the symphony could be cut from some 2 hr to about 20 min and the intermission eliminated.

Conductor's concern

The conductor naturally expressed some concern and wondered if audiences would dwindle drastically.

If that happened, the consultant said, parts of the concert hall could be shut down, giving major savings in utilities, janitorial services, and wages.

In the worst case, if nobody came, the concert hall cafe could be opened daily and the orchestra sent to work there.

Keep it simple revisited

Jim West **February 13, 1995**

A long time ago a column in this space issued a plea to writers of all sorts — journalists, English majors, engineers, whoever — to keep their writing simple.

At that time, Oil & Gas Journal endorsed the best advice a writer can have: Write to inform, not impress.

Now, more than 20 years later, it's time to thump the tub again for writing devoid of wordy phrases that get in the way of communicating a fact or opinion.

Useless words

A steady flow of press releases from oil companies and government agencies around the world shows a remarkable mix of good and bad writing. Included in the latter category are releases from writers intent either on proving they know lots of words or making doubly sure no one misunderstands their story. The symptom is overuse of adjectives and adverbs that could best be cut from the story.

A classic example, one we have used before, is a press release covering a substance the writer called "molten liquid sulfur." Molten or liquid alone would have done nicely. There was no need for overkill.

There are many other examples of useful words that are useless in their context. Among those that readily come to mind are imported foreign oil, manufactured polyester, offshore Gulf of Mexico leases, future growth prospects, past experience, uniquely different, and positive benefits.

It's a rare day when newspapers or newscasters don't report something was completely destroyed or partially damaged. Or maybe it completely disappeared. And don't forget to look or listen for advertisers' offers of a free gift.

Too many words

Once in a while a press release comes across our news desk that is refreshingly simple even though the subject is tough to explain to shareholders.

Such a release from a producer might say simply, "Our profits are down because oil and gas prices, which are beyond our control, are down. The culprit is a surplus of energy."

Instead, writers and management often seek refuge in wordy releases that mean the same thing but try to sound supersophisticated in doing so. In those cases we read, "Our profitability was negatively impacted due to the fact that an oversupply situation depressed prices for oil and gas, our primary source of income."

An oversupply situation, of course, is the opposite of lack of availability, which in simpler terms could be called a shortage.

Substituting short words for long ones is easy, once you get the hang of it. For example, initiated litigation could be replaced by filed suit, entered into an agreement means agreed, a consortium is a group or combine, at that point in time can be shortened to then, and in the near future means soon.

All this is not to say that other writers make all the mistakes while we at Oil & Gas Journal make none. A close reader can soon spot our mistakes. Some of them are dandies that have us muttering, "How did that get in there?"

We try to avoid such mutterings — sometimes successfully, sometimes not.

Hello 486/50!

Keith Rappold **February 20, 1995**

Oil & Gas Journal is officially joining the acronym crowd by converting from a mini-mainframe dedicated word processing system to personal computers (PCs).

The worst part about this switch is every editor's bane, the undefined acronym.

CAD, RAD, RAM, ROM. What looks like a word puzzle is really just a small sample of undefined terms casually used in computer manuals, magazines, and help forums. Sure, everybody knows what those acronyms mean because PCs have been around for years. Or do they? We had been using a perfectly fine (until now) system that did not have a new acronym each week.

The largest problems with the dedicated system were a lack of speed and poor compatibility with standard PC files used by the rest of the petroleum industry.

These are problems no more! Now we can use articles and files in Word, WordPerfect, WordStar, or word whatever.

Our desks now sport top-of-the-line machines with fancy Windows-based software and all kinds of bells and whistles. These PCs should provide many years of editing pleasure.

The mainframe word processor lasted 15 years. Many publications, especially newspapers, still use that system.

Another one bites the dust

In recent weeks, OGJ editors have taken the plunge and rid their offices of the old monochrome dumb terminals and huge keyboards. The hallways and store room have become a graveyard of wires, bulky keyboards, and monitors with strange parts and cathode-ray tubes reminiscent of horror movies from the 1950s.

We will surely miss that old keyboard, which was laid out specifically for editors and writers. Gone are the helpful dedicated keys such as Delete Character, Delete Word, Delete Line, and Delete Block (my favorite). The sturdy 8-lb keyboard was twice as thick as a PC keyboard and had a large heat vent to cool the circuit boards and warm one's hands in winter. The keyboard looked as if it were designed for use by roughnecks offshore.

The hour-long wait for the mainframe to come on line each day caused many an editor to bald prematurely.

And warm too! The mainframe unit could easily heat a small village. Soon, the heavy 3 ft X 3 ft X 4 ft mainframe box with its twin 80-megabyte hard disk packs (each about the size and weight of two college calculus books) will return to its maker. The outdated monitors will not be so lucky.

The changeover to the new system started last year and will be completed any day now, once all the files are transferred and the last dumb terminal unplugged.

Will the readers notice the change? We surely hope not.

Rumorbusters

Anne K. Rhodes **March 20, 1995**

Recent events have created more uncertainty in the U.S. Environmental Protection Agency's reformulated gasoline program than has existed at any time in the program's dubious history.

EPA exempted Alaska from the program because motorists there reported headaches, nausea, and other health problems thought to be related to use of RFG. EPA also postponed compliance with the RFG program for 39 counties in New York, Pennsylvania, and Maine.

In addition, New Jersey plans to opt out of the program next winter. And now Milwaukee residents are registering health complaints similar to those experienced in Alaska.

Recently, a photo in the Houston Chronicle showed a Waukesha, Wis., resident holding a sign that read "The EPA's new gas exploded my Bronco engine."

Although Oil & Gas Journal cannot make any pronouncements about the toxicity of MTBE, we can squelch the rumor that the additive, and others like it, can cause engine explosions.

Peroxide formation

OGJ recently received a letter asking whether ethers used in RFG can form peroxides — very reactive compounds that can create explosion hazards. The reader was concerned about the two pairs of unshared electrons on the oxygen atom that separate the two ends of the ether molecule.

An industry source tells us methyl tertiary ethers such as MTBE and TAME have a very low tendency to form hydroperoxides because there are no secondary or tertiary hydrogens on the carbon atom next to the oxygen. The other side of the molecule (the methyl side) is "saturated" and therefore stable.

On the other hand, ethyl and isobutyl ethers have a higher tendency to form these peroxides, but not much more so than olefins, which have been a gasoline component for decades. In the case of ETBE, this increased peroxide-forming tendency comes from the ethyl side of the molecule, not the tert-butyl side.

In gasoline, hydroperoxides initiate polymerization, which forms gums. Refiners add antioxidant packages to gasolines to reduce gum formation and stabilize any peroxides that have formed.

But even without stabilization, these chemicals would not become explosive, says our source. Instead, the peroxides would cause refiners difficulty in meeting gum specifications for gasoline.

The peroxides also bite into gasoline octane. Active oxygen levels of 20-30 ppm in gasoline can knock off 1 or 2 octane numbers.

Preaching to the choir

Although this information may not be news to many refiners, a little "preaching to the choir" can't hurt: The new, reformulated gasoline is no more likely to cause an engine explosion than is conventional gasoline.

Now let's get to work educating the public.

Restructuring as a way of life

Jim West **April 24, 1995**

Some things never end.

Take petroleum industry restructuring, for example. ("Please. Take it," as a famed comedian would say.)

The restructuring movement that simply won't go away brings to mind the so-called gas bubble in the U.S. Or, more accurately, in the words of one industry executive, "It's not a bubble. It's a sausage."

Just when you thought there was an end to restructuring, downsizing, rightsizing, or whatever you call it, another company unveils plans for things such as layoffs, office closures, sales of assets, or a merger.

Hard times

Behind all those corporate spasms is a steady diet of hard times.

The American Petroleum Institute spelled it out clearly when it reported employment in the U.S. upstream sector has sunk to its lowest monthly level in 20 years. API based its estimate on figures from the U.S. Bureau of Labor Statistics.

API said about 331,000 persons were employed upstream last January, the smallest number since May 1975 when 325,000 were working. About 423,500 U.S. upstream jobs have been lost since February 1982 when monthly employment in that sector stood at its peak: 754,500.

The culprit is a laundry list of things that are familiar to anyone who follows the industry: puny oil and gas prices, higher costs because of environmental rules, higher taxes, and lack of access to prime federal acreage.

Those factors and others haunt all segments of the industry.

That is apparent in 1994 annual reports explaining the reasons for bad news like reduced earnings. In a prime example of forthrightness, Consolidated Natural Gas Co., Pittsburgh, began its report to shareholders: "1994 was not a good year for CNG." That's good communication.

More to come

When will the restructuring movement end? Never, predicts Watson Wyatt Worldwide, a Washington consulting firm.

Driven by customer demands and fierce competition, companies around the world expect to continue — even increase — their restructuring programs well into the future.

"Reorganizations are no longer isolated events that companies go through and put behind them," said David MacCoy, a Watson Wyatt consultant. "Constant change is here to stay."

His firm surveyed more than 1,800 CEOs, CFOs, and senior human resources managers in Canada, France, Germany, Japan, the U.K., and U.S. to reach that conclusion. The survey covered industries that employ more than 18 million persons, or more than 6% of the working population in those countries. The oil and gas industry was included, of course.

More than 90% of survey respondents said their companies went through some sort of restructuring within the past 2 years — but with mixed results. Although about half of the companies achieved cost cutting goals, far fewer hit key operational targets in productivity and customer satisfaction.

That's an ominous finding. It proves what rank and file employees have long suspected: Any bizwhiz in management can cut costs. The trick is to do so without destroying your company's production and customer bases.

With those bases gone, management's cost cutting problem will vanish. A defunct company has no costs left to cut.

Red beans, rice, and tabouli

L. R. Aalund **May 8, 1995**

There is an often torrid place in the desert on the shores of the Persian Gulf that offers a range of ethnic cuisine that almost rivals that available in New York, London, or Houston.

You want Japanese sushi, Arabian tabouli, Italian lasagna? You got it. Cooks there also prepare Pakistani, American, Indian, Sri Lankan, Filipino, and Northern European food. Many could walk into a big city ethnic restaurant and go right to work.

However, in this case cooks are feeding over 5,000 workers in separate mess halls serving, either a la carte or cafeteria style, some 15 different types of ethnic food.

The workers are building a huge LNG complex and port at Ras Laffan, Qatar. Chiyoda Corp. is the main contractor, which accounts for the Japanese delicacies. The diversity of the rest of the food program reflects the many ethnic groups in the work force.

Worldwide catering

Directing this food operation is the French company Sodexho, a management services company. It runs catering and food operations around the world. As examples, it handles food operations in hospitals, universities, and on offshore rigs, operates an intermission bar at the Helsinki City Theater, and caters for such diverse events as an elegant ball in Wiesbaden and the Albertville winter Olympic games.

It recently passed Marriott as the world's largest caterer, according to Akram Itani, commercial manager for Sodexho's Qatar operations. He manned a booth at the Middle East Oil Show last March in Bahrain.

He explained to this editor that the Sodexho affiliate in Qatar not only cooks for the Ras Laffan project but handles cleaning, security, and maintenance for it and staffs the clinic.

But this editor kept the discussion centered on two of his favorite subjects: food and the oil industry.

Sodexho will offer as many different menus and get the specialized cooks for them as the client wants, but it generally is not feasible to have separate kitchens for only a few expatriates of the same nationality. Workers can change mess halls or "cross eat" if the client wants. But with hundreds involved, that could create havoc.

Food and productivity

Ethnic food is a critical element in project productivity, Itani says. First there is a culture shock when a foreign worker is placed in an alien environment. "Home cooking" helps overcome the shock. Then there sometimes are religious restrictions that must be considered.

Finally, Itani says, the body reacts to foods and spices it is acquainted with. It cannot function well without them.

At the moment, there are no soul food or Tex-Mex kitchens at Ras Laffan, but Sodexho is supplying a workboat operating in Qatari waters that has its own cook who serves red beans, rice, and other Cajun standbys. There's little doubt where this crew, or at least the skipper, is from.

A 137 year old right-of-way

Warren R. True **October 9, 1995**

Within a few weeks, Diamond Shamrock will commission its 405 mile, 10-in. pipeline to move 25,000 b/d of products from its McKee, Tex., refinery near Amarillo across Southeast New Mexico to El Paso.

In the last few miles, the line intersects history when its right-of-way follows about 30 miles of the rocky, dusty track of the first U.S. mail service from the Mississippi River to California: the Butterfield stage route.

Southern route

John Butterfield and some partners won the contract after the U.S. Congress in 1857 authorized mail service to the West. The Overland Mail Co. was to deliver mail between St. Louis and San Francisco over a route that covered 2,795 miles.

The more northerly route of the Santa Fe Trail was rejected for one that traversed Missouri, Indian Territory (known today as Oklahoma), Texas, the

New Mexico Territory (today's New Mexico and Arizona), and on to San Francisco via Los Angeles.

The part of the Butterfield route partially followed and crossed by Diamond Shamrock's line runs along the horizontal border between Texas and New Mexico, cuts north of present day El Paso, runs through Mesilla near Las Cruces, then cuts westward. El Paso was the Butterfield route's midpoint.

Butterfield crews took only 1 year to survey the route, prepare the roadbed, locate watering sites, and build bridges, transfer stations, passenger stations, and corrals.

More than 800 employees were hired, 1,500 horses and mules bought and distributed, and 250 coaches and mail wagons bought. When the first stage left St. Louis in September 1858, Butterfield, age 56, went along for the first leg and was reported "to know every foot of the ground and to be known to everybody."

(It is not recorded whether this founder of American Express carried his card.)

The first mail from California arrived in St. Louis Oct. 9, 1858 — 137 years before publication of this week's Oil & Gas Journal.

Generally, stations were 20 miles apart, with the longest stretch between stations being the 52 miles between Mesilla and Cooke's Spring. Stopping only to change horses, a passenger could expect the journey to take roughly 3 weeks. For that, he paid $200.

Postage for a letter cost a hefty 10¢. Freight rate was $1/100 lb/100 miles.

New rates?

The Butterfield route operated from 1858 to 1861 when outbreak of the Civil War forced what was left of the U.S. to use a route from Missouri via Salt Lake City and Sacramento. The Butterfield route was not used again.

Presently, refined products will begin filling a new tank farm in El Paso, awaiting local distribution, shipment to western U.S. markets, or possible movement into Mexico.

It's not known how Diamond Shamrock's rates compare with those of Overland Mail Co.

Zeroing in on ZEVs

Bob Williams **December 11, 1995**

A long war of words over electric vehicle mandates in California has come to a cease-fire.

The auto and oil industries have been eyeball to eyeball with state regulators and environmental lobbyists for 5 years over the state's zero emission vehicle (ZEV) rule.

Last month, one side blinked.

Headed for a "train wreck?"

California Air Resources Board in mid-November ordered its staff to develop a new proposal by March 1996 to change its 1990 rule that 2% of all new vehicles offered for sale in the state in 1998, or about 20,000, must be ZEVs — in effect, electric vehicles.

With a new rule, the board then can cut automakers' ZEV quotas or delay the mandate.

CARB is backing off its 1998 rule for a simple reason: Not enough people are sufficiently atwitter about the idea of buying a $50,000 golf cart that can go a whopping 120 miles at a time at a sizzling top speed of 50 mph before having to be recharged all night, fueled by a 1 ton, $10,000 battery that has to be replaced every couple of years.

In other words, a mandate to make these vehicles available is useless unless somebody wants to buy them.

A team of scientists assessing advanced battery technology at CARB's behest confirmed what the auto industry has been saying all along: The technology just isn't there yet.

One CARB board member, an engineer (How did he slip in there?), said the ZEV program is "headed for a train wreck" if the 1998/2% mandate is enforced.

Automakers reportedly have approached CARB with offers to introduce ZEVs commercially as early as 1996 — but in volumes considerably less than the 1998 mandate.

CARB insists it is not backing off its plans to cut emissions as planned, which would require widespread introduction of ZEVs. The agency also is sticking to its fleet ZEV quotas of 5% in 2000 and 10% in 2003.

ZEV risks

Automakers are concerned not only about the risks of developing commercial ZEVs that run head on into consumer indifference but also about the pace of battery technology progress leapfrogging to the point that early ZEVs are quickly obsolete.

With automakers still delaying investments needed for large scale production of ZEVs, there's speculation in some quarters CARB will either have to extend the production deadline still further or provide significant incentives for automakers to accelerate their plans.

Oil companies oppose such incentives as unfair subsidies to automakers and essentially a hidden tax on consumers. Refiners are irritated because they invested heavily—without subsidies—to produce cleaner gasoline in keeping with CARB deadlines.

Electric vehicles aren't new. They've been around since the dawn of the auto industry.

They never took off for the same reason they're floundering now, all the mandates by the People's Republic of California and millions of automakers' R&D dollars notwithstanding: Not enough buyers. They're being bought only by Hollywood millionaires trying to prove their ecologically correct credentials.

Here's a fearless prediction: the number of electric vehicles sold in the U.S. in 2000 to average, middle income households at a price that reflects the vehicle's true cost: zero.

R+M/hamburgers

L. R. Aalund **December 18, 1995**

We cover the refining industry like a blanket but rarely report on the retail service station business. This sector has integrated oil company top management ecstatic. It's not because of gasoline but because of roaring sales of snacks, beer, even lettuce.

France's Total seems to have made the best study of what food, drink, and other items Europeans want at service stations. It is building perhaps the world's largest on about 9 acres in Luxembourg. It will have 2,200 sq m of commercial space.

Some frequent these service station/food marts because they can't find a parking place in town. In Germany, antiquated early evening and weekend store closing laws often force working singles and couples out on the autobahn to service stations with less restricted hours to shop for their groceries.

A Shell station operator in Germany reports his food operation kept him from folding during the Shell Brent spar boycott that swept Europe this year.

Germans go for good bread. Esso AG, Exxon Corp.'s German subsidiary, offers baked goods in 200 of its more than 400 filling station/food marts.

Snacks and drinks account for one third the income at such stations, Esso says.

Made in the U.S.A.

The service station/convenience store concept was invented in the U.S., where companies are pushing the envelope with it.

Star Enterprise plans to welcome customers at an Atlanta station with a 1.5 min. video on station specials while they tank up. A Star manager says this turns "dead time" into "productive information time."

"Fun time" at the pump would also seem to be a concept worth pursuing. Pumps could be equipped with internal slot machines and a fourth readout window so the playing customer could see whether three cherries pop up to win.

War of giants?

Exxon Co. U.S.A. didn't respond to a question about how many food marts it has. Such stonewalling inspires wild scenarios in the mind of an idle editor who has spotted cheeseburgers in cold cases at Exxon and Chevron stations in Baytown, Tex., a refining citadel. Rather limp affairs, but undeniably cheeseburgers.

Is there a secret agreement in which Big International Oil has agreed not to sell hamburgers if Big International Burgers doesn't sell gasoline? History hints this is the case. Do the Baytown cheeseburgers mean the surviving Seven Sisters are getting ready to put on aprons and start serving hamburgers?

That could be a mistake.

Think about it. McDonald's, Wendy's, Burger King, and others all have very efficient drive-up operations. Put a gasoline pump by the takeout window and thousands of times a day it could be, "A hamburger, order of fries, diet Coke, and fill 'er up." Pumps with golden arch livery would look classy.

Marketing experts would probably come up with catchy names for new entries in the burger/gasoline market that would associate the newcomers with famous brands in sandwiches or fuel.

Perhaps we should stop this speculation now — a half column earlier, you say?

Just remember, though, you read it here first.